Hottest of the Hotspots

Critical Green Engagements

Investigating the Green Economy and Its Alternatives

James Igoe, Molly Doane, Tracey Heatherington, Melissa Checker,
José E. Martinez-Reyes, and Mary Mostafanezhad
SERIES EDITORS

Hottest of the Hotspots

The Rise of Eco-precarious Conservation Labor in Madagascar

Benjamin Neimark

THE UNIVERSITY OF
ARIZONA PRESS
TUCSON

The University of Arizona Press
www.uapress.arizona.edu

We respectfully acknowledge the University of Arizona is on the land and territories of Indigenous peoples. Today, Arizona is home to twenty-two federally recognized tribes, with Tucson being home to the O'odham and the Yaqui. The University strives to build sustainable relationships with sovereign Native Nations and Indigenous communities through education offerings, partnerships, and community service.

ISBN-13: 978-0-8165-4238-3 (hardcover)
ISBN-13: 978-0-8165-5664-9 (paperback)
ISBN-13: 978-0-8165-5196-5 (ebook)

Cover design by Leigh McDonald
Cover photograph by Garth Cripps
Typeset by Sara Thaxton in 10/14 Warnock Pro with Trade Gothic Next LT Pro
and Baskerville URW

Library of Congress Cataloging-in-Publication Data
Names: Neimark, Benjamin, 1972– author.
Title: Hottest of the hotspots : the rise of eco-precarious conservation labor in
    Madagascar / Benjamin Neimark.
Other titles: Critical green engagements.
Description: Tucson : University of Arizona Press, 2023. | Series: Critical green engagements:
    investigating the green economy and its alternatives | Includes bibliographical references
    and index.
Identifiers: LCCN 2023003678 (print) | LCCN 2023003679 (ebook) | ISBN 9780816542383
    (hardcover) | ISBN 9780816551965 (ebook)
Subjects: LCSH: Biodiversity conservation—Economic aspects—Madagascar. | Nature
    conservation—Economic aspects—Madagascar. | Biopiracy—Madagascar.
Classification: LCC QH77.M28 N45 2023 (print) | LCC QH77.M28 (ebook) | DDC
    333.7209691—dc23/eng/20230614
LC record available at https://lccn.loc.gov/2023003678
LC ebook record available at https://lccn.loc.gov/2023003679

Printed in the United States of America
♾ This paper meets the requirements of ANSI/NISO Z39.98-1992 (Permanence of Paper).

Contents

Illustrations

Figures

Table

Acknowledgments

To my mother and late father and my family, Celine, Raphael, and Cerise. Without their support, this book would never have been written.

This book, if anything, is a love letter to a fascinating and unique place. It is a very small way to give back to a country and its people who have always been so generous with their time, and overly patient with me, especially as I fumbled at trying to speak Malagasy and endlessly bothered them with questions about their life.

My only hope is that the book can be used in some way to shed light on the livelihood difficulties of rural Malagasy and their attempts at navigating sometimes very intrusive conservation and development schemes. I am also hopeful that some of the voices reflected in this work from Malagasy scientists, field guides, local conservation and development workers, and others less visible, will assist in illuminating the uniqueness of the importance of Malagasy nature and the plight of those whose livelihoods depend on it.

The downside of writing a book that should have come out a long time ago is that I now cannot fit, or remember, all the names of colleagues and friends who have helped and supported me throughout this journey. For this, I am sorry to those I have not added to this very abridged list below.

This work would not have been possible without the leadership and extreme professionalism of the following Malagasy colleagues who caried out such great research in the face of very difficult conditions, including Lahiko Charlotte Nagnisaha, Pierrette Razafindravelo Miza, Herimino Manoa Rajaonarivelo Nohary Fitiavana Ramanarivo, and Mialitiana Rabemananoro. Malagasy family and friends who over the years who have supported my work include Michel Ratsimbason, Lalasoa Ranarivelo, Tatamo Michèle Ralambomanana, Eric Randrasana, and Jean Claude Ratsimivony.

I also want to thank those co-authors who while working together have provided direct insights into my work. These names include Richard Schroeder, Bradley Wilson, Sango Mahanty, Wolfram Dressler, Tim Healy, Patrick Bigger, Oliver Belcher, Jacob Phelps, Catharine Corson, Laura Tilghman, Christian Kull, Saskia Vermeylen, David Amuzu, Sarah Osterhoudt, Lloyd Blum, Christina Hicks, Camilla Toulmin, and Simon Batterbury.

There are those who have facilitated this work with intense and fruitful discussions, including Karolina Follis and Luca Follis, James Fraser, Jeffrey Newman, Nigel Clark, Frances Clever, Jared Margulies, Ivan Scales, Mez Baker-Médard, Jacques Pollini, Mary Mostafanezhad, James Igoe, Amber Huff, Julia Jones, John Childs, Bram Büscher, Rob Fletcher, Glenn and Tori Lines, Giovanni Bettini, Kevin St. Martin, Genese Sodikoff, David McDermott Hughes, Tor Benjaminsen, Hanne Svarstad, Brendan Coolsaet, Rosleen Duffy, Dan Brockington, Sian Sullivan, Steve Goodman, Julie Hanta Razafimanahaka, Rosanna Carver, Gwyn Campbell, James Simon, Sarah Laird, Chanelle Adams, Beau Pritchett, and Dillion Mahoney.

Again, there is no way I am remembering all those who have helped along the way, but I thank you as well.

Hottest of the Hotspots

Malagasy botanist making herbarium specimens for drug discovery in Northern Madagascar. (Photo by author)

Introduction

The Green, Blue, and Bio-economy and the Rise of the Eco-precariat in Madagascar

Hotspot Discourse: Making Space for Conservation Commodities

In the late 1980s, researchers published a series of articles that drastically modified the global conservation map. Calling attention to locations with unusually high concentrations of species endemism or those found nowhere else on earth, and areas facing exceptional threats of species extinction, they argued that these "hotspots" should be accorded the highest priority for protection.[1] The original article identified ten hotspots for protection. Two years later, the list was expanded to eighteen. By the year 2000, it had grown to twenty-five, and by 2004, thirty-four hotspots had been proposed for special attention.[2] Yet throughout this period of hotspot proliferation, one site in particular—the island of Madagascar—was continually recognized as one of the "hottest" of all the hotspots[3].

It is easy to see conservationists' attraction to Madagascar. Formed from the large supercontinent Gondwana over a hundred million years ago, it is the world's fourth-largest island.[4] The country's large size and distinct topography, as well as its remoteness from the African continent, provides a unique environment for this unparalleled biodiversity to flourish.[5] According to the theory of "natural selection," when large islands like Madagascar separated from larger landmasses, similar species on each developed selective

Figure 1 Vezo woman planting mangrove propagules as part of a blue carbon conservation project in Madagascar. (Photo by Garth Cripps)

adaptations as they encountered new localized environmental pressures. Over time, the species that stayed on Madagascar took on unique adaptations that differed from their now distant cousins.

It is no exaggeration to say that Madagascar is a true tropical treasure, teeming with unprecedented levels of species richness and unparalleled endemism.[6] These include over a hundred known species of lemur, nine species of massive baobabs (or "upside down" trees), and countless wild succulents found across its one-of-a-kind "dry forests." With few parallels in terms of species richness, Madagascar boasts over fourteen thousand vascular plants, of which up to 90 percent are endemic. Furthermore, 50 percent of the island's birds and 98 percent of its amphibians, reptiles, and mammals are also found only on the island and nowhere else.[7]

Yet the country's exceptional rates of endemism also have a catch. Endemic species have higher potential threats of extinction, and as the thinking goes, if these are lost from Madagascar, then they are lost to the world. It is this extreme endemic biodiversity and extinction threat relationship that has led many conservation projects to specifically target Madagascar as a place in dire need of protection.[8]

In this vein, Madagascar sometimes goes by another moniker: the "Great Red Island." The name is attributed to Madagascar's overexposed red laterite soils.[9] Satellite imagery of the red soils, as presented in popular literature, depicts a drastic and sometimes misleading snapshot of a denuded wasteland.[10] The cultural anthropologist Genesee Sodikoff states that even from Madagascar's earliest days, natural historians and visitors viewed the country "as a damaged or wounded island," needing outside intervention to save its natural wealth.[11]

Over the years there have been countless attempts by the state, and now civil society and private sector actors, to ward off environmental degradation, albeit with mixed results.[12] Madagascar continually ranks as one of the world's hottest hotspots and is targeted with some of the most aggressive fundraising for conservation found anywhere on earth. For the global conservation community, if ever there was a biodiversity crisis, Madagascar is its ground zero.[13] According to the thinking of some of the main architects behind the hotspot strategy, endemic island countries like Madagascar are the best sites for getting the "maximum bang for your conservation buck."[14]

It is not only conservationists who have taken note of the value of Madagascar's fascinating biodiversity, however. Plant parts and insects are extracted out of the high, humid forests of the east; succulents are gathered in the western dry-spiny forests; and soft coral sponges and microorganisms are found on the northern reefs. The unique flora and fauna have distinctive biological traits and exceptional chemical properties, highly attractive for commercialization into new natural products, including drugs, crops, chemicals, and biofuels. It is in this context that the commercial sector, like hotspot conservationists, place a special value on Madagascar's nature.

The systematic search, screening, collecting, and commercial development of valuable genetic and biological resources is called "bioprospecting."[15] As the term suggests, bioprospectors, like those who search underground for gold or semiprecious stones, are on an exploration mission—to isolate the distinctive chemical scaffolding transformed by years of evolutionary history. Madagascar's plants have a long history of bioprospecting. A chance discovery of an anticancer agent taken from the indigenous Malagasy flower rosy periwinkle in the United States by the pharmaceutical giant Eli Lilly helped treat millions of children with childhood leukemia and others with non-Hodgkin's lymphoma. This discovery netted the company hundreds of millions of dollars in revenue. Although the original plant was not collected in Madagascar, it is thought to be an indigenous species, but no revenues from the discovery were ever returned to the island, prompting criticism and leading it to be referred to as the first ever case of "biopiracy," or theft of nature and knowledge.[16]

The classic image of bioprospectors is embodied by descriptions of the daring western ethnobotanist surrounded by "indigenous" forest dwellers, preparing a concoction of a "traditional" medicinal remedy.[17] However, since the onset of new, advanced technologies in the life sciences, genomics, and biotechnology, it is not only the lone scientist searching for new drugs, but large foreign and Malagasy firms, many of whom are associated with highly structured conservation projects orchestrating global commodity chains of biogenetic material. Bioprospecting in Madagascar begins with the collection of tons of leaves and bark from the most remote villages, brought to large-scale pharmaceutical labs in the Global North and run through super-high-throughput bioassays capable of a million screens per day. What was once the provenance of the "barefoot doctor," looking for the cure to all of humanity's ills under the canopy of the rainforest, is now an elaborate industrial process that uses some of the most advanced technology and capital finance to turn *nature into drugs*.

In theory, the commercialization of nature from bioprospecting should provide monetary and nonmonetary compensation back to local resource users and national governments for "market conservation" and to combat species extinction, climate change, and deforestation. Advocates claim that the production of new natural products and services will help generate market access, employment, and skills, and upgrade infrastructure in high-biodiversity countries, taking advantage of the abundant and underutilized natural resources and accessible cheap labor and land. However, in reality, turning nature into "conservation commodities"—defined as natural products and ecosystem services for market conservation in Madagascar and across similar sites of high biodiversity in the Global South—takes place under significant tension.

Madagascar is one of the world's poorest countries, and smallholders, who constitute up to 80 percent of the country's rural population, are targeted as the main beneficiaries of these economic interventions; yet many live on less than $1.50 a day.[18] The growth of market conservation and development interventions (henceforth, market conservation) raises critical moral economic questions surrounding their legitimacy and ability to improve livelihood strategies for those adversely affected by the interventions them-selves, particularly their capacity to compete in the global marketplace and navigate the messy entanglements associated with it.[19] Here lies the para-dox, namely conservation's now overall reliance on the market—for many, the main culprit in ecological degradation—to fix the environmental crisis of its own making.[20] Having started out as piecemeal, small-scale income-generation interventions at the 1992 Earth Summit in Rio, now the breadth of nature's commoditization is all-inclusive, from the air we breathe to the soil microbes under our feet, and even in deep-sea thermal vents.

This book takes an original approach to the contemporary history and politics of market conservation and development initiatives in the Global South under the lens of ecological crisis. In particular, it follows three dif-ferent contemporary market conservation interventions: the green, blue, and bio-economies. Green economy interventions, such as the practice of biodiversity offsetting, are defined as the practice of, for example, large-scale mining or aviation industries compensating for their environmental damage by funding conservation or making payments for ecosystem ser-vices. The latter acknowledge existing contributions of ecosystems through monetary or in-kind payments to countries or local communities with high biodiversity. The closely associated bio-economy incorporates nature-based discoveries, such as new drugs, and industrial products, such as biofuels, in the life sciences, biotechnology, and engineering, in an attempt to decouple from fossil fuels and other "dirty" industries. Last are blue economy pro-grams, which promote marine-oriented growth, from mangrove "blue" car-bon sequestration to the extraction of deep-sea minerals used in renewable storage devices, such as lithium-ion batteries. Woven into the logic of the UN Sustainable Development Goals (SDGs), this "triple nexus" of green, blue, and bio-economy–based market conservation now shapes the contours of national environment and development strategies throughout the devel-oping and industrialized world, and are the main drivers to commoditize nature for purposes of saving it.[21]

Making nature into commodities under the green, blue, and bio-economy has proven to be particularly hard work, well beyond what conservation and development planners imagined. Yet much of this work still goes unnoticed, particularly in local areas of Madagascar. I examine the political economy of market conservation through the work that goes into transforming nature into something tradable and the effects of these interventions on those most vulnerable—many of which are advocated by conservationists and development planners as humanity's last resort to save nature and, quite possibly, humanity itself.

As geographer Heather Lovell notes, "You can't value what you cannot measure."[22] Building on this, I argue that you cannot measure without the knowledge and manual labor to do so. This is what this book is about: it is a story of the historical and contemporary attempts to make nature markets decipherable for market conservation, or what geographers Patrick Bigger and Morgan Robinson refer to as the process of making "nature legible" for commodity production under different forms of environmental crisis.[23] I specifically focus on the generally understudied labor force behind market-based conservation in the green, blue, and bio-economy. In the following sections of the book, I question: Who are the rural Malagasy who get recruited into this work; how do they live; and why does this matter so much for conservation?

Whose labor makes nature "legible to capitalism?"[24] I, alongside colleagues Sango Mahanty, Wolfram Dressler, and Christina Hicks, developed a framework to describe the dichotomy between the skilled "eco-profician" and unskilled "eco-precarious" workforces necessary to the functioning of market conservation, and to better understand the complex and uneven relationship between market conservation and types of labor embedded into these frontier economies.[25] Put simply, we describe the proficians as a transnational managerial group—technicians, project consultants, and managers who make up the visible spectrum of labor. This group is highly skilled and most likely to benefit from an expanding green economy—what Standing calls the "global proficians."[26] In our framework, we establish three categories that form an emerging body of "eco-precarious" laborers who are respectively professionalized casual workers, hyperprecariat, and dispossessed.[27] These heterogeneous groups encompass a diversity of workers who are incorporated into the green economy platforms in different ways—some voluntarily and others out of desperation. Yet most of this work remains hidden under the veneer of participation and inclusion in the green, blue, and bio-economy discourse.

Using a lens of eco-profician/precariat labor helps "to highlight not only those who lose/gain opportunities for paid work" through market conservation but also "whose labour becomes and can remain informal and therefore precarious."[28] Results show that many times it is this latter group of eco-precariat who find themselves working for, and against, their own survival in

Figure 2 Local Malagasy hired to measure mangrove carbon sequestration potential. (Photo by Garth Cripps)

targeted sites of conservation, and at times essentially *monitoring their own livelihood demise*. If anything, this is a story about how even the best projects can fail to achieve conservation goals yet succeed in other objectives that have much longer-lasting impacts. The approach taken in this book offers a window into the role of precarious labor in doing the work to make nature into commodities within periods of global ecological crisis, including climate change, deforestation, and biodiversity extinction.[29] In doing so, the book critically examines different perspectives across the "green" value chain—at the local, regional, and international scale[30]—from rural smallholder Malagasy farmers and local participants to large industry representatives, research scientists, and policymakers in the United States, UK, and Europe. Rather than sequestered and static interventions, this book treats these economic development models as dynamic. It explores the history of market-led conservation and sustainable development with a case study of bioprospecting (chapter 3), and more contemporary empirical chapters on biodiversity offsetting and blue carbon (chapter 4 and 5, respectively). While local participation is front and center in the conservation and development discourse surrounding the green, blue, and bio-economy, the book foregrounds the role of skilled and unskilled scientific labor in making these projects. It borrows from feminist political ecology, highlighting the gendered complexities of uneven benefits and burdens around access and control of valuable natural resources, and the privileged knowledge of their use by more powerful groups.[31] It is this lens of labor, where privileged access to resources and capital of powerful policy groups and national elites crystalizes, that is my point of departure, highlighting both the positive and negative implications in terms of rights, equity, and sustainability for those upstream actors in the green commodity chain. As the oft-used saying goes, "Madagascar is where development dreams go to die."[32] Grounded in over fifteen years of multisite ethnographic fieldwork, this book examines the contours and significance of market conservation and development. Contrary to the above, it is relatively optimistic in its outlook. My approach is critical, yet pragmatic. It is written for audiences within and beyond academia, providing a balanced view of the messy entanglements of labor, science, business, conservation, and development, and has wider implications for the practice that previous research has, to some degree, overlooked.

Structure of the Book

The first chapter, "Never Let an Ecological Crisis Go to Waste," introduces the book's main theoretical points of departure, re-historicizing the complex social relations of development across the green value chain. It presents the concept of "commodity frontiers," or the multiple ways those leading conservation and development programs expand capitalism's reach, appropriating new natures through market interventions.[33] It helps to build a deeper understanding for readers of the book to see the central paradox, or the reliance on natural commodities and services under market logics to solve multiple social and environmental crises. The chapter discusses the complex use of crisis narratives of climate change, deforestation, and biodiversity extinction in the blue and green economy discourse, and the justifying of nature's survival through, and for, extending and deepening market interventions. It sets the scene to understand the analytical approach taken to examining work that goes into making new environmental products and services, while exposing the visible expert labor and the many-times hidden eco-precarious work that make market conservation and development possible.

Chapter 2, "From Kings to Conservationists: Crisis, History, and the Biodiversity Hotspot Frontier," provides a background to the geographical land-

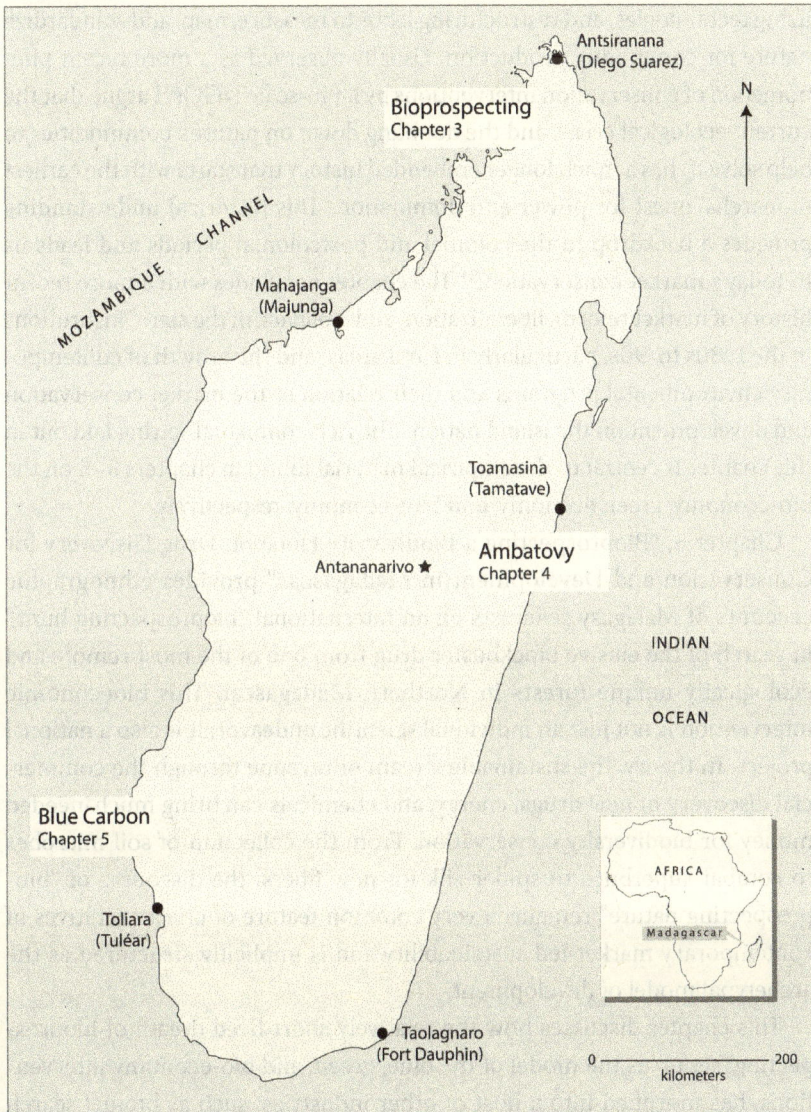

Figure 3 Locational map of three case studies in book. (Author WFH)

scape of global biodiversity hotspots, particularly one of the world's hottest hotspots—Madagascar. The chapter surveys Madagascar's political, sociocultural, and political-economic history, and more broadly explores how it fits within the larger global environmental governance of green value chains. It demonstrates that there is a long history of constructing institutions, mobilizing technologies, and restructuring labor to measure, map, and standardize nature for commodity production. Usually observed as a more recent phenomenon of conservation interventions by large-scale NGOs, I argue that the current ecological crisis, and the doubling down on nature's commodities to help solve it, has a much longer embedded history that starts with the earliest monarchs' quest for power and domination. This historical understanding provides a backdrop to the colonial and postcolonial periods and leads us to today's market conservation.[34] The chapter concludes with a more recent history of market reform, liberalization, and "rollback of the state" institutions in the 1980s to '90s, particularly in rural areas, and the growth of contemporary environmental programs and their relation to the market conservation and development on the island nation. The rich contextual setting laid out in this chapter is central to the empirical material found in chapters 3–5 on the bio-economy, green economy, and blue economy, respectively.

Chapter 3, "Bioprospecting a Biodiversity Hotspot: Drug Discovery for Conservation and Development in Madagascar," provides ethnographic accounts of Malagasy scientists on an international "bioprospecting hunt," in search of the elusive blockbuster drug from one of the most remote and ecologically unique forests in Northern Madagascar. This bioeconomic intervention is not just an individual scientific endeavor; it is also a national project. In theory, the sustainable stream of revenue through the commercial discovery of new drugs, energy, and chemicals can bring much-needed money for biodiversity conservation. From the collection of soil microbes to combat superbugs to spider silk for new fibers, the discourse of "bioprospecting nature" remains a very common feature of crisis narratives of contemporary market-led sustainability and is implicitly structured as the archetypal model of development.

This chapter discusses how the relatively short-lived dream of bioprospecting, set up as the model of the blue, green, and bio-economy interventions, has morphed into a host of other industries, such as biofuel search and production, and how the labor of collecting this biodata has set up the offsets, leading to the focus, in chapter 4, on biodiversity offsetting. I introduce what might be the future of bioprospecting, through biomimicry and extremophiles in the deepest darkest corners of nature—from deep-sea trenches to outer space, a theme that is picked up in the conclusion.

Biodiversity offsetting is the practice of reducing the net loss of biodiversity, caused by intensive mining and other extractive industries, through the protection and mitigation of intact ecosystems. In chapter 4, "Green

Extractives: Biodiversity Offsets and the 'License to Trash,'" I highlight how biodiversity offsetting, particularly for large-scale mining projects, has become a key conservation strategy in Madagascar, where corporate mining interests work to reshape their image as environmental stewards, both locally and globally. Biodiversity offsetting is a practice that enables nature to be traded under what many have come to call the "economy of repair."[35] Offsetting signals a shift in the way localized environmental externalities of market transactions are remediated through the production of an environmental good or service removed both spatially and temporally from the "affected" or degraded nature, or as a way to make environmental damage pay for its own survival through enterprising conservation and development. This chapter displays perspectives from key architects of offsetting. It discusses the labor that goes into the ordering of nature needed to exchange it for net losses or net gains of biodiversity. New offset spaces (including conservation enclosures and repaired sites) are intentionally designated and designed based on biodiversity thresholds and metrics created by experts, who have created, maintained, and mined the rich biodata produced through earlier iterations of collection, carefully monitored by auditing institutions, many of which fall short of delivering on their development goals at the local level.

The last of the empirical chapters, chapter 5, "Extending the Frontier: Blue Carbon and the Commodification of the Sea," is about the blue economy and discusses how the oceans have become a new expansion of green growth strategies into coastal and deep-sea marine space. It runs the spectrum from conservation of important protected areas to the mineral extraction of deep-sea trenches. Still taking shape, the nascent blue economy agenda can be observed as a way to navigate the pitfalls and shortcomings of land-based growth of the green economy and circumvent the sticky issues of landed property rights, and especially local customary claims to land and forests, for those local and indigenous people who rely on access to it for their livelihoods. Therefore, a "shift to the blue" can be seen as a somewhat purposeful move "away from the green." This move is not without its own set of problems, as these two conflicting and at times opposing agendas—one from conservation and a second from industry—seek to redefine the oceans in the way that fulfils an economic imaginary of these frontier extractive zones.

The conclusion of the book, "Reconsidering Conservation Commodities," illustrates the principal implications of understanding the practice of frontier economies in a period of global planetary crisis through the lens of smallholders. It identifies the paradox at the center of market-based sustainable development: that many of the attempts to commodify nature have not only fallen short of their expectations to save it but have also upended the sustainable development goals these programs set forth to accomplish. We return to bioprospecting in this chapter and discuss how it is used to justify access to, and conservation of, deep-sea extremophiles and, more commonly, corals, organisms promoted in the "blue economy" discourse as uncharted frontiers holding the key to novel drug discovery and other bio-discoveries. Running through much of the preceding analysis is the idea that somewhere, hidden in nature, is the key to our environmental problems, if we can just find it. This "silver-bullet strategy," using the tools of capitalism, represents the ambivalence central to sustainable development overall, and its central paradox. The chapter culminates with a review of some of the possibilities and shortcomings of postgrowth alternatives, as they relate to alternative development in the Global South and market-based conservation and development. It provides a snapshot of perspectives from Malagasy scientists on the future of market conservation in Madagascar and how they see their role in shaping it. The book concludes with a call for scholars to broaden the scope and utility of the eco-precariat framework beyond conservation and development in the Global South to include the precariousness and hyper-precariousness of workers in energy transitions, adaptation work, and an array of green economy labor in the Global North.

This book is targeted at an interdisciplinary group of scholars from fields as diverse as anthropology, environmental studies, geography, marketing, science and technology studies, political science, and sociology who are concerned with sustainable development. I try my best to write in a no-jargon style, although I admit this has been more challenging than expected. It includes some deep theoretical and contextually rich empirical material to capture the interest of students at both the undergraduate and postgraduate levels, as well as the broader public who will find the book engaging and, most importantly, readable. In all my writing, I have sought to maintain analytical rigor while presenting material in a clear, accessible manner that should appeal to educated lay readers who are interested in sustainable development. While the book focuses primarily on sustainable development initiatives originating from the United States and Europe, its analysis has implications for understanding the practice throughout the Global South. Hence, I hope the book will appeal to an international audience, providing the foundation for similar research in other contexts.

Chapter 1

Never Let an Ecological Crisis Go to Waste

Nature, Markets, and the Eco-precariat

On October 14, 2019, the seventy-seven-year-old Emeritus Rabbi, Jeffrey Newman, was arrested outside the Bank of England, while protesting in support of the climate activist group Extinction Rebellion.[1] An online video of the arrest in *The Guardian* shows the Rabbi, in a full white prayer shawl and yarmulke with an embroidered ER symbol, being carried off by two uniformed police officers.[2] A seasoned environmental campaigner, the distinguished Rabbi, in a statement to the press, said that "we are in a period of enormous, catastrophic breakdown, and if it takes an arrest to try to find ways to galvanize public opinion, then it is certainly worth being arrested." According to the Rabbi, "What I want to say is that Extinction Rebellion . . . is activism, but underneath it's also about rebuilding, about showing that a society can function better when people collaborate." Interrupted before he could finish, the Rabbi told officers surrounding him that he disagreed with what they were doing and did not accept their grounds for his arrest. "It's not OK!" he shouted at the arresting officer before being taken away.

Extinction Rebellion was targeting the Bank of England with the intention of "disrupting the system bankrolling the environmental crisis. . . . The ecological damage is global, and it is hitting the Global South now." Protesters said they were switching their focus to the financial institutions "funding environmental destruction."[3] Since 2015, Extinction Rebellion has garnered worldwide attention with the teen climate activist Greta Thunberg and global

Figure 4 Climate activist Rabbi Newman arrested outside the Bank of England during an Extinction Rebellion protest in 2019. (Photo by Jill Mead / *The Guardian*)

student protests. As Thunberg puts it: "Think we should be at school? Today's climate strike is the biggest lesson of all."[4]

On the surface, it might seem that Rabbi Newman's protests on behalf of ER were not in vain. Only three months later, Larry Fink, the founder and CEO of BlackRock, one of the world's largest investment firms that manages roughly $9.6 trillion in global assets,[5] sent shock waves through the global finance community with a public disclosure that the company would completely overhaul its investment decisions toward "environmental sustainability."[6] Fink stated that climate change "has now put us on the edge of a fundamental reshaping of finance," and that "companies, investors, and governments must prepare for a significant reallocation of capital," which was said to have "marked a watershed moment in climate history."[7]

The announcement by Fink, not coincidently one week before the start of the Fiftieth Annual World Economic Forum (WEF) in Davos, was welcomed by many in the environmental community, as it was essentially seen to end big finance's relationship to coal investing, and quite possibly the "end of coal itself."[8] Notwithstanding, Fink's reaction stands in stark contrast to Rabbi Newman's plea for financial institutions *to step aside* and make room for a global climate movement, which in his words, includes "activism," "participation," and "rebuilding."[9] In fact, Fink is making a very different case in response to the climate crisis, one that proposes deepening and intensifying capital's relationship to nature, through sustainable financing, impact investing, green bonds, capital risk management, and, in effect, *much more, not less, market capitalism.*

According to the late Neil Smith, the 1980s and '90s kicked off a watershed moment for this brand of concentrated financial solution to the growing environmental crisis. In this era, characterized by Thatcher-Reagan's "market triumphalism," emerged a spectrum of what Smith called "ecological commodities," which advocated a market-based environmentalism and thought that nature could help pay for its own survival through new finance-backed conservation and development schemes.[10] For Smith, this was "nothing less than a major strategy for ecological commodification, marketization, and financialization which radically intensifies and deepens the penetration of nature by capital." He went on to say:

> A new frontier in the production of nature has rapidly opened up, namely a vertical integration of nature into capital. This involves not just the production of nature "all the way down," but its simultaneous financialisation "all the way up." Capital is no longer content simply to plunder an available nature but rather increasingly moves to produce an inherently social nature as the basis for new sectors of production and accumulation.[11]

Feminist geographer Cindi Katz echoes this, arguing that this period was not just another attempt by capital to exploit nature, but a transformative shift in nature's "status and temporality," including the rollout of financially infused "representations of nature," such as biodiversity and wetland banks, natural capital, carbon credits, green bonds, natural assets, and ecosystem services.[12] According to some critics, this period moved the needle radically in how society was beginning to understand, and act on, nature under capitalism.[13] Critical geographer Jessica Dempsey explains that this period signaled an "enterprising of nature," where nature in its many forms could now "pay for its own survival" through new forms of investment. Even as we speak, novel "smart" conservation programs roll off the assembly line, including platform capitalism and blockchain and Bitcoin schemes, which seek to both deepen and widen capital's reach.[14]

Naomi Klein's book *The Shock Doctrine* provides a point of departure for how economic openings arise in a particular period of environmental crises.[15] In Klein's first chapter, "Blank Is Beautiful," she describes the business opportunities afforded to corporations under postdisaster conditions of chaos and disorder. She states that under "disaster capitalism," or "the brutal tactic of using the public's disorientation following a collective shock—wars, coups, terrorist attacks, market crashes, or natural disasters," lay the groundwork "to push through radical pro-corporate measures."[16] Klein links this dynamic of disaster as an economic opportunity for "shock therapy" and discusses how these measures are by no means new, but their scale of operations are nothing less than awesome. Klein provides diverse examples of corporate-humanitarianism (post-2003 Iraq, the 2004 Asian Tsunami, and Hurricane Katrina) that were once the domain of government agencies and state institutions, and exposes the true nature of being under a "disaster capitalism blueprint." Recently, she has reiterated her thesis of disaster capitalism in discussing the recent bailout under the COVID-19 crisis, climate change, and its repercussions for the planet and its most precarious inhabitants.[17]

Similarly, in Madagascar and elsewhere in the Global South, ecological crises, such as climate change, deforestation, and biodiversity extinction, have recently helped mobilize capital's expansion into new economic frontiers—from the forest floor to the deep seas. Yet, recently, critical scholarship has engaged the "green" frontier economics with appropriate skepticism. From the fringes to the mainstream, different forms of conservation commodities have transformed our relationship and understanding of how the market and nature interact under crisis. This deepening and widening of commodification under green capitalism includes the identification and subsequent valuation of nature into less tangible and more fungible services, biodiversity and carbon offsets, and credit-based "natural capital."[18] As the empirical cases in this book on offsetting and bioprospecting in forest and marine areas in Madagascar demonstrate, even as green financialization advances, there remain large-scale investments in conventional commodities for green development worldwide, such as food, fibers, forests, fashion, and fuels. Indeed, running parallel to green financialization, states, financial institutions, multinationals, and national elites continue to grab resource stocks (e.g., land, timber, mines) as an accumulation strategy.[19] Producing these traditional "hard commodities" while also developing new financial products and services under the banner of the green, blue, and bio-economy thus enables these investors to speculate and acquire future claims on resources, to hedge against risk, and to finance bond and other financial service markets.[20]

According to historical geographer Jason Moore, "planetary crisis" is exactly what happens when capitalism intensifies and expands. For Moore, and others historically, capital expresses itself through the state "to map, identify, quantify, and otherwise make nature legible to capital."[21] This process is essential to reorganize nature in a way that is intelligible for commodification and financialization. Nevertheless, to (re)organize "nature" is by no means a simple process; indeed, it is a herculean task. Christian Parenti shows us that it is possible, but only through the mobilization of what he calls "geopower," or the "technologies of power that make territory and the biosphere accessible, legible, knowable, and utilizable."[22] And as Robertson notes, "[u]nder capitalism, however, these technologies of measurement and abstraction are used specifically to define adequate bearers of value."[23] However, as I demonstrate in this book, while the entwining of the private sector, science, governments, and global capital may collectively all be drivers behind the grand project of making nature legible for market conservation, there are a host of local scientific elites, and other skilled and unskilled workers, who are actually doing the hard work (e.g., counting and monitoring mangroves, collecting plants species, and collecting soil carbon) with mixed results for not only the projects but also their own livelihoods.

While many "green" products and services represent an "economy of appearances"[24] that may never fully materialize into any form of recognizable market,[25] their performative nature still has diverse impacts on livelihoods.[26] These effects have exacerbated tensions over who can gain, maintain, and control access to resources, or whose labor is appropriated and in what ways.[27] What is the tangled role of "labor" in making commodities legible and investable? In this book I not only contribute to the ways in which nature is actually being commodified, if indeed it is, but also extend our conversation to include labor politics in the green, blue, and bio-economy. More generally, throughout the three empirical chapters, I ask what a green, blue, and bio-economy labor force looks like, how commodities within these economies are produced, and what the emerging landscape of labor politics is that is helping to produce them.

There has been a litany of scholarship attempts to better understand the fundamental role of institutions and organizations in "making nature legible"

or defining the ideological shifts of market-led environmental and conservation policy within the burgeoning green, blue, and bio-economies. Much of
this scholarly work identifies "neoliberal" or market-led discourse and institutional performance and practices, which facilitate access to biodiversity and
the sharing of benefits and burdens that derive from nature's commercialization,[28] as well as the evolution of conservation funding to ease resistance to
corporate capital.[29] Yet throughout this work, less attention is given to those
low-paid workers who are charged with the day-to-day tasks of collecting
the data that fill up the calculative databases. Following Mackenzie's claim
that to say that "economics is performative is to argue that it *does* things,"[30] I
ask, beyond constructing powerful economic discourse, how do markets in
the green, blue, and bio-economies perform? In a more material sense, who
exactly is doing their "dirty work" or counting and identifying species and
hauling them through the mud back to base camp, because, in the end, it is
through these labor tasks that new drugs and high-quality biodiversity and
carbon offsets are made real, thereby providing the scientific legitimacy and
social currency for market conservation to function.

Conservation Commodities and the Green Value Chain as a "Lived Experience"

Geographers James Fairhead and Melissa Leach note that seeing the conservation and development "landscape metaphorically" as "text" or "spectacle" allows us to view "discursive fields linked to particular institutions [or
groups], and how these, in turn help to shape socio-cultural processes" and
the material effects on local environments and livelihoods.[31] Market conservation is therefore not only a top-down or externally driven process carried
out by large-scale NGOs and elites, but also a lived experience permeating
society at different regional and local scales. For example, Peet and Watts
highlight the potential of local "environmental imaginaries," or rather "whole
complexes of imaginaries," that can enhance, inspire, and give institutional
strength to resist through new social and political movements.[32] While these
"local imaginaries" are powerful as a counterhegemonic form of resistance,
others remind us that these "discourses from below" can be easily co-opted
and reshaped by development actors and used to the advantage of more powerful groups.[33] Recently, there have been more studies examining the socio-

environmental relations surrounding the green, blue, and bio-economies within Madagascar, which up to this point have generally been examined with a focus on the performance of state institutions and nonstate actors and their effects on rural communities, but with much less focus on the local labor force, whom the programs themselves are designed to benefit.[34]

I examine market conservation through the lens of political ecology to see how imagined green, blue, and bio-economies and local ecosystems interact. The production of conservation commodities, much like previous historical and contemporary boom commodity crops, such as rubber, cotton, and sugar, restructure "geographic space at the margins of the system in such a way as to require further expansion."[35] For instance, organizing wage laborers and the restructuring of customary land tenure arrangements are shaping, and are being shaped by, the promotion of new markets globally, and the increased push to commodify new forms of nature, such as biofuels, carbon offsets, and other natural products.

This book describes what happens when market conservation touches down on the ground in Madagascar. The next chapter explains the historical and contemporary mobilization and restructuring of a precarious workforce whose purpose is to make the island's biodiversity-rich nature into a usable resource for development. Initially this was about the island's early rulers exerting control over the territory and its inhabitants. In Madagascar's more recent history, however, rural labor has been used to roll out new market conservation and development streams under the green, blue, and bio-economy. It is to this history that we now turn.

Chapter 2

From Kings to Conservationists

Crisis, History, and the Biodiversity Hotspot Frontier

> The gulf between land tenure facts on paper and facts on the ground is probably greatest at moments of social turmoil and revolt.
>
> —Jim Scott, *Seeing Like a State*

> The business of protecting species and habitats, the "intellectual labor" of foreign experts—some who stay a short while in Madagascar, and others who make the island their permanent home—has overshadowed the contributions of subaltern labor. This category consists of Malagasy porters, builders, servants, and guides, who for a small wage have made life livable for foreign travelers and who have enabled the conditions of possibility for scientists to discover new species in Madagascar's ecosystems.
>
> —Genese Sodikoff, *Forest and Labor in Madagascar*

Seen from the outside, the island nation of Madagascar has always suffered from some form of environmental crisis. Early in the eighteenth century, explorers and visitors to the island were already raising the alarm over perceived local pressures put on its rare flora and fauna.[1] The loss of its biodiversity, of course, was not the only emergency thought to besiege the island nation. Malaria, foreign invasion, and long-standing political fragility put pressure on Madagascar's leaders to leverage its natural resources for commodity extraction in attempts to preserve power and instill a means of social control.

The historical residue of turning the country's rich resources—everything from slaves to sapphires, tropical hardwoods to marine microbes—into something tradeable, is something that still shapes the commodity landscape in Madagascar, remaining a mainstay of development planners to this day.[2] This historical relationship between commodity production and development has been uneven, serving those in power, who are sometimes very far from the shores of the island itself. Yet the consequences of relying on the island's endowed resources (i.e., on nature) to pull it out of crises have been far-reaching, having significant effects on Malagasy society and the island's environment, particularly at the local level. The anthropologist Andrew Walsh termed this "Madagascar's Global Bazaar," where the island's natural resources and labor are "systematically valued and devalued" at different points in the commodity chain.[3]

For many, this may seem like a familiar narrative of the oft-cited "resource curse"—a poor African country, rich in resources and exploited for profit by the West.[4] However, the story of power and control of Malagasy nature is not so straightforward. As I demonstrate below, years of wrangling for political control over the massive island and attempts to resist exploitation by colonial or foreign powers have produced sometimes perverse development outcomes for the island nation. Moreover, nature in Madagascar, and elsewhere for the matter, has been shown to be difficult to transform into commodities.[5] Nature's values are diverse; property rights over resources are often contested; and biodiversity is intrinsically difficult to convert into the interchangeable units that enable standardized goods or services to be easily recognizable for market exchange.[6]

Natural resource commodity production in Madagascar also faces other challenges. A large chunk of the country is filled with nutrient-poor red laterite soils, making large-scale agriculture inefficient; its forests are challenging to traverse, and known mineral resources are sited in remote and difficult-to-access areas. The country also historically lacks quality infrastructure, such as good internal roads and ports, and its distance from its historic trading partners in Africa and Europe makes it less amenable to global exchange.[7] Lastly, the country's rare, exceptional biodiversity is almost impossible to replicate, and replication is a prerequisite for commodity production.

To solve this problem, a host of new institutions, technologies, and labor, especially scientific labor, have been restructured to measure, map, and standardize nature for commodity production. This chapter explores the systematic mobilization of both skilled and unskilled labor to take scientific calculations and create the market tools to turn nature into commodities in Madagascar. It builds on what James Scott refers to as the practice of "forging tools of legibility," made possible through calculative advances in scientific forestry and cartography, laying the framework for control over property rights and revenue derived from its natural resources.[8] In Madagascar, such acts of legibility are constructed from historically appropriated scientific labor and local knowledge. I argue that this practice of making nature legible, carried out by the monarchy, colonial, and independent state, and now the nonstate and private sector, is one of modern control over Malagasy forests and agricultural lands, and moreover, the "unruly" labor needed to produce them. This provides a background to one of the world's "hottest biodiversity hotspots" and to the emergence of commodities as part of the development of value chains and the contemporary green, blue, and bio-economic platforms in sustainable development. I examine the history of calculation through the introduction of modern market tools, weights and scales and measurements, the adoption of coinage, systematic plant collectors, cartography, and scientific forestry.

Jason Moore argues that modern capitalism, and the ecological crisis that followed, did not start in the nineteenth-century industrial revolution smokestacks of England, but long before, in the sixteenth-century sugar plantations of the Caribbean and the subjection of its people.[9] The chapter also shows that the antecedents of market conservation in Madagascar didn't necessarily begin with the growth of modern civil society NGOs and private sector–aligned donor institutions, but rather with the earliest monarchs, carried through into the colonial and postcolonial independent state. My purpose of this chapter is not to rehearse the insightful work done previously by colleagues on the history and politics with the rise of market conservation and development in Madagascar.[10] Rather, I seek to highlight how contemporary market tools are built on such historical legacies shaping the Malagasy landscape for conservation commodities, or making biodiversity, land, labor, and markets legible for commodification under crisis. The chapter concludes by highlighting some of the more contemporary large-scale modern projects of mapping mangroves, forests, agricultural land, and geological resources, which in effect make contemporary environments of Madagascar amenable for the market.

Who Are the Rural Malagasy?

It is important to note that although there is a distinct national identity, the ethnic population of Madagascar is very heterogeneous and loosely divided into twenty ethnic groups who all speak different dialects of the same Indio-Malayan language.[11] There are also Arabic, Swahili, and Bantu language influences in the coastal regions.[12] While there have been periods of serious tension, violence, and civil strife, the ethnic groups have generally coexisted and have a history of intermixing. Yet for the most part, they maintain cultural distinctiveness, exhibited through different practices of land use, livelihood strategies, and ceremonial practices of celebration and burial.[13]

Although there are larger urban populations, both on the coasts and in the highlands, the Malagasy are largely an agrarian society. Farming is the mainstay of the economy, with roughly 70 percent of Malagasy living in rural areas. Agriculture, fishing, and forestry make up about one-third of the country's Gross Domestic Product (GDP).[14] There are diverse agricultural practices on the island. Rice production is carried out in wet rice paddies, providing the main food source, and swidden farming (called *tavy* or *hatsake*) and midrange grassland farming (*tanety*) also take place. Spices such as vanilla and cloves are grown in multistory home gardens (*tanimboly*) in the northeast, and Zebu (a local breed of cattle, *Bos taurus indicus*) husbandry occurs in the south and west of the island.[15] Fishing, which is a major source of food security for coastal communities, is generally less incorporated into national economic statistics of GDP. However, it remains a hidden pool of economic wealth, supporting the livelihoods of about 1.5 million people and contributing up to one-fifth of inhabitants' protein diets.[16]

However, rural life in Madagascar is extremely challenging, and the country remains one of the poorest in the world. Official accounts still place Madagascar at roughly 158 on the Human Development Index, with approximately 75 percent of the population of 24.2 million people living on less than $1.90 per day, and with most rural Malagasy living in extreme poverty.[17]

The extreme rural poverty in Madagascar has many causes. However, the country's agricultural output has remained a significant constraint throughout its history. Most of the island's soils have toxic concentrations of aluminum and iron oxide, which bind to critical nutrients like nitrogen and phosphorous, which are essential, but now unavailable, for crops to use. There are fertile areas in the country, including all limestone areas, bottom land, and alluvial plains of Lake Alaotra, volcanic areas like Lake Itasy, and some forest soils in the central highlands. The latter can be mobilized using swidden agriculture, improved fallows, terracing, and manure to build up the fertility of poor soils. However, the lack of available nutrients means that, overall, agriculture in the country is difficult. In fact, one might even say that Madagascar holds some of the most challenging soils in the world for large-scale agricultural production.

Over time, the island's diverse populations have developed strategies to overcome these biophysical constraints in soil nutrients, growing food including unique cash crops, and carrying out animal husbandry, which contribute to a rich agro-cultural heritage.[18] However, it is the practice of growing and eating rice, particularly lowland rice paddies and upland hill rice (*tavy* in Madagascar), which is said to hold the country together both culturally and socially.[19]

Rice is the main subsistence crop for many in Madagascar. For those who can afford it, rice is eaten three times a day, and many Malagasy will tell you that all meals must include rice, and if not, it isn't considered a full meal.[20] Rice accounts for approximately 44 percent of land under cultivation and nearly half of all caloric intake (FAO). According to the International Rice Institute (IRRI), roughly ten million people are said to derive income from the rice sector.[21] Researchers' estimates suggest that most farmers cannot produce enough rice to feed their families.[22] Seventy percent of Malagasy grow rice, yet an estimated 67 percent are net rice-buyers and 80 percent are reported to have bought rice at some time during the year.[23]

Tavy is a traditional practice of shifting cultivation, used to grow rice in the uplands and overcome the soil fertility challenges. *Tavy* utilizes the burning of forest vegetation to release nutrients to produce upland rice and is thus a vital part of Malagasy food security. This system can be followed by a two-year cycle of maize and beans, sweet potato, and/or cassava crops, with the number of harvests and length of cultivation cycle depending on soil capabilities and climate. When this agricultural cycle has run its course after two or three years, fields are left fallow (*savoka*), allowing a return of secondary growth.[24] The regeneration of natural vegetation is a traditional practice for restoring fertility to the system. As a result of years of mixed-land use, clearing and frequent burnings for agriculture, plantation forestry, and hunting, land in Madagascar is now covered by pockets of primary forest and secondary vegetation (such as grasses and ferns), as well as areas of invasive shrubs and trees containing pockets of primary and dense secondary forest.[25]

Shifting cultivation is a practice widespread in the high-humid and lowland forests of the eastern forest corridor, and this has been an intricate part of the Malagasy agricultural landscape for many centuries. It is particularly important for those who lack access to lowland wet paddies or mechanized irrigation and provides a lifeline to overcome times of food insecurity, particularly during the lean months or what many in Madagascar call "the bridging period" (*periode de soudure*) between rice harvests. Rapid population growth, mixed with insecure land tenure and lack of food security, has led to a shorting of fallow periods from fifteen years to five years. This reduces yield due to a reduction in organic matter, causing some to extend production into new primary forest zones. There is usually a rapid decline in fertility when there is no suitable forest left to maintain the growing population.

For years, *tavy* has been criminalized and outlawed, but it remains a persistent practice deeply rooted in both the culture and history of the highland Malagasy farmer. *Tavy* rice is an intricate part of the many rituals and sacrifices that lie at the heart of Malagasy rural society,[26] customs handed down from farmers' ancestors, and the *tavy* field is a sacred site where the farmers can directly communicate with their families' descendants.[27] The custom has also been associated with political resistance by peasants during the colonial era, and serves as an organizational framework for village life, defining gender roles and individual responsibilities.[28] Fires have been observed to increase during election years: the clear historical marginalization and political and economic disconnect between the mainly coastal Betsimisaraka and highland Merina ethnic groups is on display during the burning season.[29] Malagasy claim customary rights to practice *tavy*, enabling them to express their culture and maintain their identity and unity.

Shifting cultivation in Madagascar has been deemed by many within the conservation and development community as a destructive and primitive practice, linking the clearing of forested land under *tavy* directly to the extinction of rare flora and fauna.[30] Yet this "degradation narrative" centered on the increasing peasant population and poor management practices did not begin with the modern conservation movement. In fact, early colonial cash crop production included coffee and later vanilla, as well as plantation forestry,[31] and even earlier tree felling by the highland monarchy took place. Each caused environmental damage, and each placed blame squarely on the shifting cultivators. Debates rage on about the deforestation of Madagascar.[32] However, recent estimates state that between 1953 and 2014, 44 percent of Madagascar's forest was lost.[33] While it is a fact that shifting cultivators have had a significant effect on land degradation and the rapid depletion of for-

ests, the social, economic, and political marginalization of shifting cultivators and the lack of any other livelihood options for them clearly contributes
to the problem.

Current scholarship has provided a much broader interpretation of this
environmental change, including a more nuanced analysis of poverty distribution, commercial resource extraction, internal migration patterns, and
division of labor.[34] Not all smallholders are economically and politically
marginal, but most have felt the brunt of years of austerity, which began
in the 1980s and '90s, leading to crippling poverty and lack of basic social
services.[35] And most notably, due to their utter dependence on local natural
resources, from forests to coastal fisheries and mangroves, they are the most
affected by conservation and development interventions. Any changes to
their access to these vital natural resources puts their survival at risk, and
therefore careful attention to smallholders and the effects of market-based
interventions on them is a key focus of this book.

Early Trade, Market Tools, and Control over the Island

Although there exists a long-standing debate among historians, anthropol-
ogists, and linguists regarding who exactly Madagascar's first inhabitants
were, most agree that Madagascar was first encountered by Polynesian settlers sometime between the fifth and eighth centuries.[36] These early travelers
are thought to have made successive stops in coastal Africa and Arabia, and
they were already active in the trade networks of the Indian Ocean.[37] Information from archaeological remains of these early settlements indicates a
vibrant coastal trade in manufactured goods and agricultural produce as
early as the tenth century, and imported goods in the fifteenth and sixteenth
centuries.[38] At this time, a number of important trading posts were later
joined by the Arab–East African "Swahili" networks of Mombasa, Zanzibar,
Lamu, and Comoros, including rubber, gold, a variety of other products,
and slaves.[39]

The Indian Ocean slave trade opened the island to the Arabic trading
networks, which were still very active when the Portuguese slave traders
landed in Madagascar in the early sixteenth century. This "European contact"
(Portuguese, followed by Dutch) shattered the island's connection with the

Swahili trading networks. However, Madagascar still remained as a junction connecting the Indian Ocean trade routes linking Africa, Asia, and Europe.[40]

The French and British made beleaguered attempts to settle Madagascar: the British in 1645–46 at St. Augustine Bay and the French in Ft. Dauphin from 1642 to 1674.[41] Most of these settlements failed to take root early on, due to resistance from coastal inhabitants against foreign incursions. In the mid- to late seventeenth century, France and England stepped in to revitalize much of the Indian Ocean's commercial activity and began reestablishing the East African Swahili trading networks, most notably the slave trade.[42] By the turn of the eighteenth century, slaves were Madagascar's commonest export, being transported to Mauritius (known as Île de France until 1810).[43]

Concurrently, both large and small kingdoms were established in the interior of Madagascar, including the Sakalava, whose control over the western coast allowed for frequent trade of slaves out of the port of Mahajanga.[44] Similarly, around the eighteenth century, the Betsimisaraka confederation formed on the eastern coast. But none was to rival the most important kingdom of the central highlands of Madagascar, the Imerina, who eventually ruled much of the country under a centralized system of government commanded by King Andrianampoinimerina (1730–1810).[45]

King Andrianampoinimerina was the first to develop sophisticated domestic and international trade networks, which he used to consolidate Merina power through structured market regulations. Andrianampoinimerina took advantage of the sporadic breaks in fighting during the Merina Civil Wars in the late 1790s to construct domestic and international commodity networks. In doing so, Andrianampoinimerina made a host of royal pronouncements that helped to establish the foundations for future markets, including civil codes, early land demarcation, and rules governing commerce.[46] It was through the civil codes where the king began to lay the groundwork for unifying, and also attempts at controlling, the different ethnic population areas across the island.[47]

The monarch also tried to enforce the adoption of standard weights and measures, which had until this point been "crude and unreliable," as well as coinage as a means of trade.[48] These attempts were to combat very inexact estimates—for example, Indian Ocean historian Gwyn Campbell notes "[what] could be carried under one arm" indicated a weight of two to three kilograms.[49] While these systems did not necessarily take off during the reign of King Andrianampoinimerina, by the late eighteenth century the use of coins was dominant throughout Madagascar, and they were being used to purchase slaves and other goods by the owners of the flourishing plantations on the Mascarenes.[50]

Although their intended purpose or geographic scope is still debated, the first documented considerations for resource protection were a set of oral proclamations made by King Andrianampoinimerina.[51] In these statements, Andrianampoinimerina stated the need to protect forests, particularly from using wood to make charcoal, and forest burning, in support of rural livelihoods, the King declaring that the forests are a public good for the poor.[52] Yet there is considerable evidence that these declarations were actually designed to consolidate power in the hands of the monarchy and quell growing unrest and threats to Merina rule.[53]

While these environmental codes are frequently considered to be Andrianampoinimerina's biggest and longest lasting achievement in control over the island, it was the regulation of the land and commodities market where the King had his greatest long-term impact.[54] This dominance was manifested by the mapping of most of the central highlands and squeezing out the country's economic potential. With this new cadastral power, the King allocated land parcels to his subalterns, in effect setting them up as fiefs, and instructed them that he would retain absolute authoritarian control over the subjects and monopolize market infrastructure.[55] The King instructed his regional nobility to compel the laborers in their fiefdoms to construct canals and drain swamps, enhancing the land's productivity. Thus, for the first time, the central highlands became agriculturally fruitful through an intricate system of wet paddy rice cultivation.[56]

These social and political changes were all part of a broader plan to modernize the Malagasy state and establish Merina dominance through a market economy. More importantly, it signaled the Merina's attempts to expand control through extensive domestic and international trading, by instituting a very sophisticated system of specialized markets that were known by the different days of the week. At this time, slaves and precious hardwoods were traded for certain metals, ceramics, and jewelry.[57] This specialized trade continued until the early eighteenth century, when Malagasy artisans began to become more skilled in the crafts of raffia, hemp, and silk production, together with small-scale artisanal gold- and silversmithing, which eventually developed into a larger and more established market economy.

Andrianampoinimerina's Kin and Merina Quasi-dominance

The nineteenth century saw the establishment of diplomatic negotiations between the British, who by this time occupied Mauritius and South Africa, and the French, who maintained Réunion Island and a post at Sainte-Marie Island off the Malagasy east coast. The highland Merina monarchy's tenuous control over most of the island's kingdoms was solidified by King Andrianampoinimerina's son, Radama the First, who was more open to contact with missionaries and political emissaries from Europe. He eventually formed an alliance with the British, who wanted him to conquer the entire island for their own benefit, which generally aligned with his own aspirations. As Radama did not have the tools necessary to maintain control, he needed Britain's help. The British abolitionists, however, banned Radama's slave export trade, and then "shortchanged" him in terms of compensation for the lost revenue. Radama realized very early on that he had made a bad deal giving unfettered British access to the island's resources and began to cut side deals with the French to build the export trade back up. Radama rejected the British alliance in around 1824–25, and formally from about 1826, when he signed a monopolistic free trade accord with French merchants. Crucially, however, he also realized he was in a crisis: he did not want either British or French imperialism and knew that both were itching to get their hands on Madagascar.

To retain independence, Radama had to act. However, he did not have a large-scale military, capital, or tools to invest; the only thing he did have access to was human capital in the form of forced labor. He therefore turned to the traditional practice of *fanompoana*, forced labor or slavery, to build a foundation for market infrastructure. This was supported by a standing army who helped him to retain access to and control over the different kingdoms' resources across the island.[58]

This direct control over labor caused enormous problems in terms of Merina relations with the Europeans, who also wanted access to Malagasy resources and workers. What followed was a political struggle, starting with Radama's successor, Queen Ravavalona I (1828–61), and her harsh policy against foreign occupation, leading to the expulsion of many of the English missionaries, most notably the London Missionary Society (LMS).[59]

This reign was followed by Radama II (1829–63) and Queen Ravava-lona II (1829–83), each of whom held a more open policy toward European society and religion, strengthening ties and placating both the French and British interests in the island. Malagasy rulers also continued to advance laws controlling the clearance of forest land while still providing access to precious hardwoods.[60] As the American political ecologist Catherine Corson notes, "Major French companies secured concessions for mineral and forest exploitation, including for exotic hardwood species such as ebony, rosewood, and sandalwood, as well as providing agricultural land to the monarchy."[61]

It was, however, the 1881 "Code of 305 Articles," proclaimed by Queen Ranavalona II, which is thought to be the first key piece of environmental legislation in Madagascar.[62] In fact, six of these articles (101 to 106) focused on forest protection and prohibited the burning and cutting down of forests. As with the earlier codes, these were designed to reinforce the queen's power over the island, subsequently shaping future colonial and postcolonial resource claims.[63]

The new codes were also clearly a way for the queen to capitalize on rural labor. Most particularly at risk were those who were mobile, such as the pastoralists and the swidden farmers. American anthropologist Jennifer Cole notes how this was particularly burdensome on the Betsimisaraka, a coastal ethnic group who were most dependent on shifting cultivation in the eastern forest.[64] Rather than giving in to the Merina demands, many Betsimisaraka took to the hinterlands deep into forests to evade administrative control.[65]

The control of labor for commodity production was only one aspect of Merina political domination. A second, more powerful tool of social control was that of taxation.[66] The idea of mandatory labor and taxation was accompanied by a ban on house building in forest plots, which, according to Cole, began the process of forced settlement for the pastoralists and shifting cultivators who, due to their nomadic lifestyle, could not be taxed in any enforceable way. Social control through forced settlement was also a way to quell insurgency, especially from competing Betsimisaraka and *Sakalava* ethnicities, resisting Merina control in the eastern and northern coastal regions.[67]

Campbell observes that Merina policies to promote economic development and imperial expansion through domestic industrialization using methods of forced labor and taxation probably resulted in much more deforestation in the highland and eastern woodlands than either foreign plantations or peasant cultivation.[68] And in the end, these market modernization efforts to open up Malagasy forests and resources were in tension with local moral economies and failed to deliver many of the economic benefits the state anticipated, resulting in prolonged economic instability. As a result of the ensuing currency crisis and power vacuum instigated by these reforms, the Merina were pushed to rely more and more on the *fanompoana*, or unpaid forced labor, and continued imports of currencies and goods. According to Campbell, ultimately, these helped push the Merina regime to the verge of bankruptcy in 1894, and eventually facilitated the French colonial takeover of Madagascar in 1895.

French Colonial State: A History of Labor and Commodity Extraction

By the time the French officially designated Madagascar as a colony, there were already well-trodden internal and external trading networks between the Merina highland capital and Paris. Cash crops, cocoa, and vanilla exported off the east coast provided a significant revenue stream, and commercial plantations were set up for a host of new export specialty crops, including the lucrative coffee and cloves.[69] This was not all, however; everything from cattle, rice, and raffia to tropical hardwoods made its way to the colonial metropolis, helping to fill French coffers.[70]

Once Madagascar became a French protectorate, the French Ministry of Agriculture sent out a group of scientific foresters to immediately evaluate the island for more efficient forest extraction and cash crop exports. It seems that the French were particularly concerned with local forest burning and access to timber and fuelwood closer to the capital, Antananarivo.[71] Their mission was, in many ways, to assess what resources could contribute to rational extraction and "organize forest product extraction, define repressive rules and create a Forest Service."[72]

Yet by the mid-1930s, the still-young colonial French forest administration was reporting widespread burning on the island.[73] And while control over people entering the forest was certainly a concern prior to French control, in the eyes of the colonial administration, a more systemic environmental legislation was urgently needed.[74] All of this, however, seemed to run parallel with Merina policies to root out those avoiding tax collection in forest hideouts and maintain a labor pool for further extraction.[75] This legis-

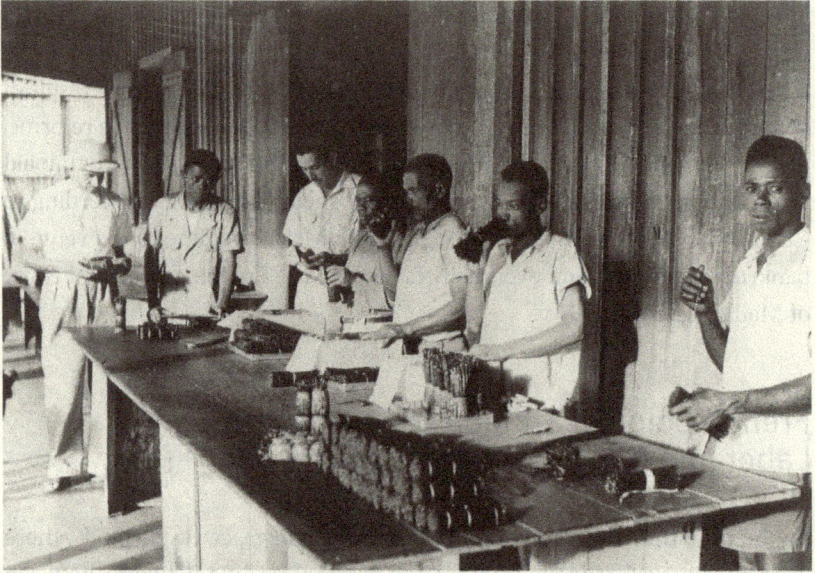

Figure 5 Colonial vanilla production circa 1952. (Photographer unknown; permission IREL—ANOM)

lation coincided with a rational scientific forestry to make sense of the "unorganized" and "messy disorder" of local systems of resource management. Forest agents brought with them modern tools and techniques expected to facilitate optimum economic efficiency, based on models developed in France. However, many of these techniques were not well adapted to the tropical humid climates and in fact failed to adapt to the economic realities of the Malagasy landscape.[76]

Natural resource scholar Jacques Pollini, drawing on work by Ramanant-soavina,[77] points to the colonial period of scientific forestry as a turning point in seeing the Malagasy labor and forests as an economic resource. Unlike the precolonial period, the top-down "command and control" approach was fueled by research field stations and the creation of a professionalized scientific labor class of foresters to work in them. This symbolized a period of foreign influence attempting economic rationalization including, at the turn of the nineteenth century, a host of reforestation programs and nursery establishments, as well as scientific trials of exotic and nontimber forest products, including rubber, and the cadastral mapping of forests.[78] These developments also included the demarcating and reordering of natural space, including, through to the mid-twentieth century, the designation of eleven forest Natural Reserves and the creation of the first two National Parks (Montagne d'Ambre and Isalo) and twenty Special Reserves.[79]

Yet this ordering of Malagasy nature did not begin with the creation of a forest service or new national park network. Even before the French takeover, there had been an ongoing and intense interest in categorizing and cataloguing Malagasy resources. Naturalists from all around the world, financed by large botanical repositories in Paris and London, were sent to retrieve economic and scientifically interesting flora and fauna to be displayed in zoos and gardens in the metropolises of Europe. There were, of course, attempts to explore, map, and collect mineral and marine resources that may also have had economic potential.

In fact, plant and animal collectors have been a fixture in Madagascar for centuries. Classification of Madagascar's diverse flora and fauna began as early as the seventeenth century.[80] Explorers, botanists, and naturalists were attracted to the island's unique wonders and imagined by some as a "living laboratory."[81] The more established European scientific expeditions, in both the eighteenth and nineteenth centuries, attracted those seeking botanical wonders found nowhere else and drawn by their economic potential back home.[82] The scientific and economic entanglement set off a mad dash, with French and British naturalists amassing large collections to classify what they saw as an increasingly diminishing resource.[83]

French naturalists Alfred Grandidier (1836–1921), Alphonse Milne-Edwards (1835–1900), Perrier de la Bâthie (1873–1958), and Jean-Henri Humbert (1887–1967), and the British naturalists Richard Baron (1847–1907) and George Shaw (1842–1917), were all instrumental in amassing huge herbarium collections in Madagascar and back in their home countries.[84] In attempts to consolidate collections and scientific activities on the island, in 1902, under the direction of the French governor Gallieni, the Malagasy Academy of Sciences was set up as the premier scientific institution in Madagascar, housing the largest botanical repository on the island.[85] This institutional collection was originally set up as a hub for European botanical gardens, such as Kew, Paris, and Linden, and ultimately had a lasting effect on how research collections would shape modern environmental interventions.[86] In fact, many of today's market-based interventions rely on knowledge set forth by the history of flora and fauna collections (see the bioprospecting and offsetting schemes in chapters 3 and 4, respectively).

More noteworthy, the amassing of these early botanical transfers of unique material was seen by historian Thomas Anderson as a "normalizing" scientific endeavor in which the island's plant material and knowledge was "universalized" to make it seem like Madagascar was more commonplace, and not in fact unique at all. Building on this, collections, at this time, were thereby justified as apolitical, free of their colonial roots under the scientific mission of saving Malagasy nature—a feature of scientific collections that persists till today.[87]

For a long time, botanical and zoological collections of usable plants and animals continued under the colonial and postcolonial era under a systemic scientific forest service. Here, forest products and tree species were housed in Madagascar, similarly to other islands in the Indian Ocean, and set up as colonial *jardins d'essai* (trial gardens)—a paradise of natural riches and oddities and for some a colonial "botanists dream."[88] These *jardins* also became trial grounds for new agricultural and forestry techniques breeding some of the most economically important species for commodity production.

Madagascar's extensive commodity development expanded with the French annexation of the island. Colonial agricultural production was designed for export, and the French divided the island into governable administrations for purposes of tax collection and the procurement of labor reserves. They also set up a system of "indirect rule" where Malagasy regional elites, foreign missionaries, and French settlers and entrepreneurs could facilitate the transport of raw materials through the major Malagasy ports located in Tamatave, Majunga, and Ft. Dauphin.

There was also a period of granting large-scale agricultural and forest concessions to French settlers and companies, mainly in the central highlands and east coast of the country. Large plantations for rice, vanilla, and forestry, as well as land for cattle production, were established in "settler freeholds" (also known as *colons*) which were run by French nationals living in Madagascar.[89] Usually established on the most fertile land, these colonial cash crop plantations put pressure on local labor reserves, pushing pastoralists and shifting cultivators onto elevated forested slopes to escape being dragged into forced labor for the colons.[90]

Exports of central importance at the time were coffee in the east, irrigated rice in the central highlands and west, and beef and maize in the west, with sisal also grown and exported.[91] Once a unique method of hand pollination of the vanilla orchid was discovered, new vanilla plantations were set up in the northeast, alongside vital cash crops of cloves, ylang-ylang, and sugar-cane.[92] Trade surpluses mounted as industrial production grew, and mineral extraction of mica and graphite maintained French economic activity through the interwar and postwar years, along with "major French companies secur[ing] concessions for mineral and forest exploitation, including for exotic hardwoods."[93] And as French influence on Madagascar increased over the course of the nineteenth century, so did their control of agricultural labor and commodity production for export.

Yet maintaining local labor was always a problem for the concessionaires. In order to guarantee the labor needed, the French brought in labor from southern Madagascar and created a tax system that facilitated the production of export crops and the imposition of a tax system for unpaid or indentured and seasonal labor, or *corvée*.[94] They also put in place a system of forced labor for various public works, which included market and road infrastructure, essentially providing free labor for export companies.[95] This pushed already precarious subsistence farmers into giving up their traditional lifestyles and providing labor for colonial export crops.

To control and manage the labor force, the colonial governments (re)imposed a per capita tax.[96] It was during this period that many of the smallholders, particularly the Betsimisaraka ethnic group living in the Eastern rainforests, suffered the most.[97] Besides the French, the ruling Merina were also beginning to encroach eastward. Betsimisaraka land was systematically divided into governable districts, each with a central fort to oversee the *corvée* labor for exportation of commodity crops, including those of rice, cattle, hardwoods, and raffia fibers.[98]

Increasing taxes, forced labor, and land pressures of commercial plantations essentially pushed more people into the forested areas and practices of shifting cultivation. The colonial administration, however, needed to make Madagascar profitable to justify keeping it. The expense of governing such a large colony far away from the metropolis necessitated the extraction of commodities, thereby transforming profit from the country's "rainforest wealth" and transforming the country's "unruly peasants" into wage-laborers.

The Malagasy themselves were not sold on such a plan. Forests became refuges from colonial forces as the Malagasy continually refused to offer up their labor to private firms and colonial officials.[99] As American anthropologist Genese Sodikoff suggests, "Betsimisaraka and Tsimihety horticulturalists and fishermen fled into the deep forest to escape taxation and coercion into industrial work sites." Eventually, stricter structural measures were insti-tuted to conscript Malagasy into colonial government–led programs, "in which conscripts ('pioneers') would serve the state for two years, and often private industrialists would 'borrow' the pioneers from public work sites for their own logging, plantation, and mining operations."[100]

This export-led development came at a price, to both the island's forests and soils. The continual need for labor, mainly from the south, for public works projects, such as building the main railways moving many of these commodities, also forced many to shift from subsistence cropping to cash crops, severely affecting food security on the island. For example, the intro-duction of coffee cultivation led to shortfalls in rice production and the extension of upland rice cultivation. Colonial pressure on the island to ramp up exports eventually led to "uneven economic development" and intensified "regional fragmentation."[101]

The tension between cash cropping and rainfed rice regimes in terms of claims on land and labor time played out in colonial agricultural settings. Due to its labor demands and attractive producer prices, coffee cultivation increased in popularity among European settlers and Malagasy farmers. As rainfed fields were abandoned (the casualties of labor shortfalls, low pro-ducer prices, cyclones, and drought), food security in the eastern region was eroded. This was also a period of advanced deforestation and environmental degradation. According to geographer Lucy Jarosz, almost three-fourths of the island's primary forests were destroyed between 1895 and 1925.[102]

The extraction demand from abroad escalated during the interwar years. Madagascar supplied castor oil, rubber, and timber to support France's bleed-ing economy. Personnel were conscripted to large-scale public works cam-paigns to work on basic infrastructure of roads, ports, and bridges, and the large investment in a new rail infrastructure also gave rise to a demand for new forest products and timber from all parts of the island.[103] Madagascar became a key supplier of forest materials to the allies during World War II.[104]

In the postwar era, the state began rapidly securing forestland, but resis-tance to French control was already beginning, culminating on March 29, 1947, with the national revolt against colonial rule. The French of course retaliated with violence, particularly against those who they saw as collud-ing against their rule, and as a result, thousands of Malagasy were tragically killed.[105] Central in the revolt were those living in the rainforests, who were said to have set the forest ablaze and reappropriated previously disposed lands.[106] Forest and resource control was central within this resistance movement against what Malgasy saw as foreign domination over their land, forest, and labor.[107]

Yet, in response to this period of unrest, the state began a process of clarifying forest boundaries and gaining more centralized control over land titling and timber permits. The state began to target those burning forests and explicitly prohibited fires in and around the forests.[108] Forest agents' powers were severely strengthened during the period, in order to enable them to enforce rules on the Malagasy living in and around forests,[109] and to enable them to protect the business interests of French concessionaires.[110] Much of the concessionaires' extraction was to return much-needed revenue back to France, which was still in tatters after the war, and having to put considerable energy into justifying its presence in Madagascar, which was becoming very expensive to maintain.

Independence: The Nation, Debt, and the Rollback of the State

After full independence in 1960, Madagascar maintained close economic and political relations with France. The links with France in this period, known as the First Republic (1960–72), under the leadership of Philibert Tsiranana, led to successive strikes and social crises for the new and fragile state. In 1975, the socialist Didier Ratsiraka gained power, promoting a nationalistic and isolationist policy.[111] This marked the Second Republic of Madagascar (1975–92), characterized by a distinct brand of social and economic policy that maintained loose ties with both the West and the Soviet Union. A number of strikes and economic crises in the early 1990s forced the coming of a Third Republic to Madagascar, marked by the rather short tenure of Albert Zafy as president (1993–96). However, after the impeachment of Zafy in 1996, Didier Ratsiraka was reinstated as president in 1997. Late in his tenure as president, Ratsiraka, following the advice of the World Bank and International Monetary Fund (IMF), instituted neoliberal reforms and open market policies that brought Madagascar into a slow-growing and uneven economic trajectory.[112] In 2001, Ratsiraka lost a turbulent reelection bid to the mayor of Antananarivo and businessman Marc Ravolomanana.

After this turbulent change, Madagascar followed its African neighbors on the path of privatization of state industries and a cutback in state-run services, paved by the World Bank and IMF in response to structural adjustment polices put in place just ten years prior.[113] In 2002 a political crisis triggered a 12 percent drop in GDP, placing 71 percent of the Malagasy population below the poverty line,[114] and the newly elected Ravolomanana first attempted to revive the national economy on a platform of economic policies of poverty reduction and corruption elimination. This brought international recognition and the return of much-needed international donor support. When Ravolomanana took office, exports of some consumer goods, such as clothing, boomed.[115] However, most of this newfound economic wealth was felt within the highland areas of Antananarivo and Antsiribe (both sites are political strongholds of the president), and the port areas of Tamatave. Furthermore, land tenure and corruption remained major challenges. After a successful second election in 2006, Ravolomanana was forced to resign in 2009, handing over power to Andry Rajoelina in what many in the West considered an illegitimate coup d'état. This caused cuts in aid to the island nation and thus starved it of vital revenue to maintain its already fragile economy. Madagascar quickly sank into a five-year political and economic crisis. After a brief hiatus, where power was ceded to elected president Hery Rajaonarimampianina, Rajoelina returned to defeat both Rajaonarimampianina and Ravolomanana and become president in 2019.

The Making of a Hotspot: From the National Environmental Action Plan (NEAP) to the Promise of Sydney

Madagascar's national interest in the protection of its environment began in 1984, when it created the National Commission of Natural Resource Conservation for Development, a body which was put in charge of preparing the Malagasy Strategy of Natural Resource Conservation for Sustainable Development, called for by some of the world's largest environmental donors and conservation organizations.[116] This body drafted a publication entitled *National Strategy for Conservation and Development*.[117] Noted as one of the first of its kind in Africa, this document reflected larger currents of biodiversity conservation seen worldwide at this time.

To launch the strategy, and garner technical assistance from the international donor community for its implementation, the Malagasy government hosted an environmental conference in 1985. The international community responded with zeal, and in 1989 helped the government of Madagascar put into practice an ambitious fifteen-year investment program known as the Madagascar National Environmental Action Plan (NEAP).[118]

However, Madagascar, like many other African countries, was faced with financial and economic crises and burgeoning debt, due to the heavy burdens of large IMF loans taken out in the 1990s. To service these debts, Structural Adjustment Policies (SAPs) were put in place by the IMF and World Bank to restructure the debt burdens. The government of Madagascar was in a precarious position, finding it difficult to accommodate the demands of international donors who supported the NEAP, but in no position to negotiate, since international donor aid was tied to the restructuring and rollback of the bloated state infrastructure. SAPs were beginning to take hold, and Madagascar's social and economic infrastructure was crumbling. Social services were cut, and the shrinking of the state through cutbacks to health care, education, agricultural services, and environmental protection was crippling the country, particularly in rural areas.[119]

Eventually, in the mid-1990s, international donors managed to bring some relief. The fast-track approval and implementation of the document into policy was a way for the government of Madagascar to obtain a reprieve from harsh austerity measures imposed by structural adjustment.[120] However, in order to obtain funds, states were pressured to adopt a host of "green conditionalities," market-based conservation strategies such as debt swaps and other biodiversity conservation interventions, which aligned with the newly established sustainable development objectives of donor institutions.[121] The influx of these funds, particularly in the 1990s, fostered a subsequent "conservation boom" in Madagascar and elsewhere in Africa.[122] This money was quite significant, with some estimates for donors' contributions being approximately $450 million for the three five-year National Environmental Action Programs (NEAP I, II, and III—1991–2008). The United States Agency for International Development (USAID) support alone totaled $123.4 million over the three NEAPs.[123] Needless to say, this money opened a window to build collaborative links with environmental NGOs and big donor projects. Furthermore, at this time, research institutes were under pressure to generate revenue through commercial opportunities.[124]

The ratification of the NEAP began under the governance of the Malagasy Third Republic and President Albert Zaffy in 1992, and was reinstated with the return of President Ratsiraka in 1997. The NEAPs were financially supported by multilateral and bilateral donors, with the goals of ensuring that the country would be able to take advantage of its unique and valuable resources to further economic development and enable its inhabitants to achieve "a better quality of life."[125]

They were, of course, aligned with the Convention on Biological Diversity's (CBD) Integrated Conservation and Development programs (ICDPs), which opened the world to both commercialization of genetic resources and climate change mitigation, which are the subjects of chapters 3 and 5 respectively.[126]

What began in the mid-1990s as a national strategy to solve environmental problems and rural development challenges has since been transformed into a full-scale conservation and development industry.[127] For example, those *cooperating* in phase one of the NEAP included donor institutions,[128] international environmental non-governmental organizations (NGOs),[129] research institutions, and land grant universities.[130] By the third phase, however, this had expanded into an environmental complex that included hundreds of NGOs, state institutions, private organizations, and multinational companies.

The country is among the largest recipients of donor aid for environmental programs in Africa, and in May 2004, the World Bank provided Madagascar with an additional $40 million,[131] a concession that was noted at that time as one of the largest ever awarded in its sixty-year history.

Following suit, in 2003 then-president Ravolomanana declared to the Fifth World Parks Congress in Durban, South Africa, that his government would triple the amount of protected area on the island nation to the IUCN-recognized standard of 10 percent of terrestrial land. It was this declaration, later known as the "Durban Vision," that helped to produce new spatial boundaries for "conservation and development" schemes to be enacted, including those of ecotourism and biological prospecting. The Durban Vision was integrated into an overarching framework known as the Madagascar Action Plan (MAP), or Madagascar Naturellement (Madagascar Naturally), developed by the government in response to the Millennium Development Goals of the UN Sustainable Development Conference of 2002. This inclusion into policy marked the correspondence of national social and political goals with the global conservation community's interests. These interests are based on years of academic and scientific research, backed by substantial foreign aid and lobbying.[132]

Roughly ten years later, at the 2014 World Parks Congress, called "the Promise of Sydney," President Hery Rajaonarimampianina announced an ambitious plan to match the Durban Vision and triple Madagascar's marine protected areas by 2024. This was to be developed under an expanding and ambitious program of Locally Managed Marine Areas (LMMA). Much like the terrestrial protected areas network developed ten years prior, local communities were to be formed into associations, or Vondron'Olona Ifotony (VOI), for comanagement and monitoring of these areas (see chapter 4 for more on the VOI). This model was built on the forest and protected areas' decentralization (GELOSE) programs and policies developed just ten years prior.[133] In many of these decentralized programs, a contract of village rules, or *dina*, would be set up in an attempt to incorporate local commitments into monitoring and protecting, in agreement with the civil society and state groups.[134] In the three following empirical chapters (3–5), we see just how the slow creep of decentralized forms of market-conservation schemes became enmeshed into the everyday of local resource users' livelihoods, with the market eventually becoming a means to organize natural resources at the local level.

Laying the Groundwork for the Green, Blue, and Bio-economy

Expanding the national park network through local participatory groups was not the only tool being developed as this time. There were several policy and development interventions that have shaped modern conservation and development into the country's current commodity frontier form. Not all of these were conceived in Madagascar, but all were either piloted or devised solely for the country. They also reflect the legibility tools, using new and innovative mapping interventions, backed up with market-based policy and infusion of capital, knowledge, tools, and techniques, essentially to open Madagascar's resources once again to investment, mostly foreign investment, through land tenure schemes, massive institutional reform of the mining sector, and finally, carbon finance.

Promoted for years by the World Bank, land tenure has been on the docket for development programs to promote economic and social development. One of the largest donor-funded land tenure projects in Madagascar in recent years was launched under the auspices of the George W. Bush Millennium Challenge Compact (MCC). In Madagascar, the MCC included $109 million for financial sector reform, land tenure, and agribusiness promotion though Foreign Direct Investment (FDI).

The thrust of land tenure reform was to harmonize potential opportunities in private investment, building on previous cartographic knowledge and the cartographic revision built on the land registry and land reform. For instance, this push for more FDI was topped off with the implementation of an investment law (No. 2007–036) and the establishment of the Economic Development Board of Madagascar (EDBM). Essentially, the EDBM was an initiative to "fast-track" foreign investment, which during its initial phase (2007–10) increased FDI inflows roughly thirteen-fold,[135] but it also had the role of administering large-scale land acquisitions for foreign-owned agrofuel production in the Ministry of Agriculture's selectively zoned Agricultural Investment Areas (AIA).[136]

Parallel to these new investment laws were an array of new land tenure laws, including a law (No. 2005–019) that declared that untitled private property was no longer under the ownership of the state, and a subsequent law (No. 2006–031) that certified long-standing customary claims to land, overturning years of centralized control.[137] Both of these policies carry immense grassroots support, mainly due to their recognition of locals' customary claims to land and the opening of municipal offices (the *guichet foncier*) to guide claimants through the process. The initiative was meant to provide tenure security at the community level through a local land certificate (*certificate foncier*), while also combatting land seizures and forced evictions.[138] These new tenure rights were loosely linked to the investment laws, in that it was thought by advocates, such as their foreign development donors, the World Bank, and the IMF, that private land ownership would stimulate economic activity, investment, and capital at the local level. However, just when these land tenure projects were due to culminate, several large land acquisitions were made public. One such deal, with the multinational Daewoo for biofuel and export grains, triggered in 2008/9 a coup d'état and five-year political and economic crisis.[139]

Second, spurred by the massive structural reforms of the mining sector, there has been a huge program to map existing and potential extractive resources in Madagascar, including a massive mapping project, which included a $40 million mission provided by the World Bank in 2007, to produce a geological and mineral deposit harmonized map of Madagascar and promote a massive institutional reform of the mining sector.[140] The multiple geological maps produced were meant to "enable the Malagasy government authorities to develop their mining activity to a level that is commensurate with the nation's geological and mining potential" and "to promote private investment initiatives, so far few in number." This large-scale project produced hundreds of digital maps, and incorporated the U.S., French, and British geological agencies, alongside the Malagasy Ministry of Mines.

There was also a massive mapping investment by the French Geological Survey (Bureau de Recherches Géologiques et Minières—BRGM), and a second with the British Geological Survey and U.S. Geological Survey. This resulted in the most detailed geological maps ever seen for Madagascar—on a 1:1,000,000 scale—a relative map of metallic substrates and industrial minerals, and most importantly, access of GIS (geographical information system) numerical databases.[141]

This work coincided with a new environment and investment law, decree no. 2004–167, or MECIE (Mise en Compatibilité des Investissements avec l'Environnement), and directives to liberalize the country's mining code (Mining Code of 2005).[142] Although long delayed, the revamp of the Malagasy mining code was pushed to revamp outdated tax policies and provide flexibility to private sector firms in a rocky landscape of fickle commodity prices and a history of overregulated and poor output by the mining industry. Besides the massive nickel and cobalt mine, Ambatovy, which accounts for roughly 30 percent of Madagascar's foreign exchange, the country has opened up to other major mining interventions, including the Rio Tinto / QIT Madagascar Minerals titanium mine, Base Resources' Toliara mineral sands project, and new graphite mines to feed a growing essential minerals market for lithium batteries (see chapter 4). However, as we will see, there has been some resistance to the mining code and its underperformance in delivering social and economic benefits to local communities.

Lastly, Madagascar has had a part in carbon finance programs, and in particular, a role in developing the infrastructure as part of Reducing Emissions from Deforestation and Degradation (REDD+). Madagascar has been at the forefront of REDD+ schemes since 2008.[143] The idea of REDD+ is that access to new conservation finance from carbon markets and mitigation trading schemes can be used to pay for projects to reduce forest clearance and degradation, which will result in lower carbon dioxide emissions. Madagascar has held five large pilot programs, with the idea of scaling up those that are successful into larger national coordinated projects, rather than "project-based" ones that allow for significant leakage or when "carbon emissions are displaced rather than avoided."[144] The total funding committed for REDD+ readiness is $10.6 million from the World Bank and Forest Carbon Partnership Facility (FCPF).[145]

In theory, if Madagascar is able to certify that over long periods it sustained protection of the forest from deforestation or degradation from the selected sites, without leading to "leakage," and that these changes are here to stay, it can trade on the voluntary carbon market after accreditation through voluntary standards.[146] However, as you can well imagine, in order to qualify for REDD+, there needs to be a significant amount of baseline understanding of carbon stocks, sequestration rates, and potential risks. In response there has been nothing less than an army of midlevel scientists and international consultants mapping and monitoring REDD+ sites. Since the onset of REDD in the early 2000's, massive geospatial analysis studies have taken place to classify deforestation using the highest resolution satellite data with land use change observations.[147] By this time, open-source software packages were available that allowed the Malagasy forests to be mapped in very fine detail. This was a game changer for REDD+ programs and opened the door for the potential real-time verification of deforestation and land use change in terrestrial forests. In fact, this technology was not only used in terrestrial forests: similar high-tech mapping technologies, such as the Google Earth Engine Mangrove Mapping Methodology (GEEMMM), were also being applied to mangroves.[148]

Two high-profile REDD+ pilots, in the protected areas of Makira and the Ankeniheny-Zahamena Corridor, run adjacent to the biodiversity off-setting sites of Ambatovy discussed in chapter 4. The REDD+ program is at the administrative core of the blue carbon mitigation discussed in chapter 5 and, as we will see, a frustration for some local resource users whose mangroves are now being commodified into blue carbon projects. Clearly, there are challenges with carbon mitigation, especially in areas where any loss of access by smallholders to forest or mangrove ecosystems risks liveli-hoods. And while there certainly seems to be significant potential in revenue generation at the state level, early lessons have shown that benefit-sharing is a substantial challenge for those who are most marginalized.[149] Research also suggests that the social safeguards meant to protect communities who depend on forests and mangroves is not yet up to speed with the implemen-tation of the projects, causing some to call for a serious evaluation of the ethics of carbon mitigation without these in place.[150]

Conclusion: Commodity Production as a Means of Social Control

Over the years, there has been a concerted effort by those in power to try to "tidy up" Malagasy nature's messy and elusive characteristics and its "unruly" rural labor force in order to make it "legible" for market conservation.[151] This process, however, is not necessarily a recent phenomenon in the way that it is generally characterized in the literature, but a long historical process of internal and external control over resources and people. As noted above, the attempts by all rulers, from the earliest monarchs to contemporary lead-ers, to politically control the massive island while simultaneously resisting exploitation by colonial or foreign powers has produced uneven develop-ment outcomes, particularly at the rural level.

Madagascar's history also demonstrates dependency conditions, where rural labor is employed in the process of commodity production, "governed" through state partnerships with contemporary market conservation organi-zations such as NGOs, private originations, and FDI. These outside orga-nizations have been able to continue and solidify their immense political and economic power through what political scientist Rosaleen Duffy calls a "governance state."[152]

For Marx, commodity production, and in particular the exploitative con-trol over labor, is inherently a violent social process, "written in the annals of mankind in letters of blood and fire."[153] Today, however, the production of conservation commodities, according to political ecologist Bram Büscher, is not written not in blood and fire, but the "golden letters" of "win-win" development narratives of inclusion and participation.[154] Couched within modernizing discourses of development and progress, this project of legibility can be viewed crucially as a way to bring land and labor into the market. Clearly, there has been a shift from the direct violence of forced labor and colonial relations, but into what? What new types of violence are now hidden in new forms of labor agreements and social relationships around the blue, green, and bio-economies?

The influence of contemporary conservation groups and their private sector allies in making and delivering environmental policy in Madagascar has been outstanding, but they do not do it alone. As I demonstrate in this book, the relationship between powerful groups of state, nonstate, and private sector actors in making nature legible for commodity extraction occurs over time and within different contexts, from the collection of plants and microbes for drug discovery to mining and blue carbon offsetting, is conducted with the support of the emerging environmental-service-based labor class.

Malagasy scientists and local-level laborers have little knowledge that their work contributes to such global programs. At first, locals were participating in projects as day laborers and wage workers in small-scale bioprospecting projects. However, as we move through time onto carbon offsetting projects for the green and blue economies, a simple change in the way forests were being managed helped spur a new mobilization of local labor. As we see in the following chapters, local groups were being employed to conduct monitoring as a way to take stock of their resources and under obligations from the local customs governing resource use, or *dina*, in a decentralized network of Community Based Associations, or COBAs. One might ask, however, when we look back on this history of structural relationships, are labor relationships in market conservation very different from those developed throughout history?

Attempts have been made to capitalize on the island's unique biodiversity and natural resources in order to address the country's chronic poverty, using ecologically friendly commodities and green markets, albeit with mixed results. I address these important questions within the rich contextual settings laid out in chapters 3–5, which explore this question as a historical case of bioprospecting, biodiversity offsetting, and mangrove carbon sequestration.

Chapter 3

Bioprospecting a Biodiversity Hotspot

Drug Discovery for Conservation and Development in Madagascar

They collect medicinal plants. Some of the plants couldn't be found in their countries so they come here. They have lots of benefits because they make the drugs in their country. . . . They'll keep them [the drugs], but we will be the ones who buy them later.

—Fidi, a local Malagasy bioprospecting worker
(Anonymous, henceforth Anon #3–1A)

There was the construction of a bridge and granary. I was there during the inauguration [of the bridge]. They were happy to see their work accomplished. The real advantage of the bridge allowed the villagers to get to the hospital easier.

—Director of the bioprospecting project in Madagascar (Anon #3–1B)

It was sometime in March 2020 when I first heard of "Covid-Organics." A news flash appeared on my TV about the discovery of a purported "Madagascar Cure" for coronavirus.[1] The COVID-19 infection rate in the UK was spiraling out of control, and everyone was desperate for any glimmer of hope to slow the spread of the virus.

Astonished, I went online and opened a BBC headline that read: "Coronavirus: Caution Urged over Madagascar's 'Herbal Cure.'"[2] Below the tagline was a picture of the young Malagasy president, Andry Rajoelina, triumphantly holding up a brownish herbal decoction (*tombavy* in Malagasy) labeled with the glossy orange and white lettering of CVO, Covid-Organics.[3]

Figure 6 The president of Madagascar, Andry Rajoelina, during the ceremonial launch of "Covid Organics," or CVO, in Antananarivo, April 2020. (Photo by RIJA-SOLO/AFP via Getty Images)

In the article, Rajoelina was quoted as stating that Covid-Organics was a "preventive and curative remedy" that "works really well." It also noted that schoolchildren were taking the tonic as a preventive measure, and orders were coming in from across Africa, with Equatorial Guinea, Guinea-Bissau, Niger, and Tanzania already receiving the first exports.[4] Astounded by the reports, I found myself wondering aloud: Could this be it? Could Malagasy scientists really have found a cure for this thing?

Immediately, Covid-Organics was met with skepticism by the international scientific community. The World Health Organization (WHO) and the U.S.-based National Institutes of Health (NIH) were already sounding the alarm about "alternative treatments" for COVID-19. It was also around the same time that former U.S. president Trump made his infamous suggestion that we might self-administer bleach as a coronavirus treatment. In a directed response to Covid-Organics, the WHO stated that they did not recommend "self-medication with any medicines as a prevention or cure for COVID-19."[5] Rajoelina's response was swift. In an interview with a French TV 24 station, he stated:

What is the problem with Covid-Organics, really? Could it be that this product comes from Africa? Could it be that it's not okay for a country like Madagascar, which is the sixty-third poorest country in the world, to have come up with [this formula] that can help save the world? What if this remedy had been discovered by a European country, instead of Madagascar? Would people doubt it so much? I don't think so.[6]

Rajoelina continued by saying:

In Madagascar, [we] have our own tonic, we are a sovereign country; we seek to help our people not become victims of the pandemic.[7] We have introduced this remedy, which contains *Malagasy medicinal plants*. This is something we are used to in Madagascar, 80 percent of the population uses herbal remedies. . . . I am just trying to tell you that African and Malagasy scientists should not be underestimated. (emphasis added)[8]

Yet the celebrated launch of Covid-Organics, even for those familiar with bioprospecting drugs from nature in Madagascar, is complicated. As noted in the introduction, the indigenous Malagasy periwinkle is one of the most well-known cases of "biopiracy"—the theft of knowledge and nature—although neither the original source material nor knowledge of making the cancer drug from it originated in Madagascar.[9]

As with the periwinkle, although there are purportedly indigenous Malagasy plants included in the tonic, the main ingredient used was *Artemisia annua*, a well-known Chinese antimalarial annual plant.[10] Many Malagasy scientists and medical clinicians also had their own response, urging caution and quietly questioning its efficiency. As one leading scientist said, "If this Chinese plant [Artemisia] is so good against coronavirus, why aren't the Chinese using it?" Others, meanwhile, were hopeful that this discovery, if effective, could put the country's scientific institutions "on the map," providing some desperately needed resources and investment.

Bioprospecting is generally defined as the systematic research discovery of drugs from nature. The practice is, in a sense, the quintessential case of the modern bioeconomy—the rush to turn nature into commodities under periods of cascading environmental crises.[11] Advocates see the bioeconomy as a "transition economy that promises to increase efficiency, optimize and (re-)value natural resources, while decreasing environmental impact through reduction in ecological degradation, waste, and greenhouse gas emissions."[12] The bioeconomy is "one of the oldest economic sectors known to humanity, and the life sciences and biotechnology are transforming it into one of the newest."[13] Its proponents argue that the switch over from industrial-style agricultural production and resource extraction to "knowledge-based" industries of biotechnology and life sciences can deliver a portfolio of bio-innovations, including alternative forms of energy (e.g., biofuels), intermediate inputs (e.g., biochemicals), and natural products (e.g., pharmaceuticals and bioplastics).[14]

For many Malagasy scientists, opportunities such as participating in international bioprospecting represent a once-in-a-lifetime chance at demonstrating their scientific relevance on the global stage. If they can somehow contribute their unique biodiversity toward the development of new natural products, then that is certainly a good thing. If Madagascar's biodiversity might help pay for conservation research, then many believe this would be even better. However, similar to other market-based conservation, the history of bioprospecting suggests that just as these programs began in the early 2000s, the industry had already moved on to another type of market-based development scheme. Payments for ecosystem services and biodiversity offsets were gaining steam (see chapters 4 and 5). Large-scale bioprospecting as a conservation tool, it seemed, had finished before it had even started. Many of the partners already knew this, but for them it was always about getting funding to collect and categorize as much Malagasy nature as they could. In this respect, modern bioprospecting has most certainly been a success, amassing some of the largest botanical collections on earth.

However, bioprospecting is by no means over as a practice, and neither has the discourse surrounding the practice disappeared; quite the contrary. From the collection of soil microbes to combat superbugs, to spider-web silk for new fibers and deep-sea marine organisms used in new drugs, bioprospecting for critically endangered species remains a very common feature of crisis narratives of contemporary market-led sustainability and is implicitly structured as the archetypal model of development. As a leading policymaker, who chose to remain anonymous, recently said to me:

> How many species there are, how scarce? Undoubtedly, we can save a lot of things [species], yet "critically endangered," as a concept, has a large amount of power, and its use is immediately apparent for environmental development and sponsors of research and conservation. There are a lot of people

out there *still* interested in this stuff, corridors of power and this crisis discourse only adds motivation for conservation.[15]

Alongside ecotourism, bioprospecting was one of the first global attempts at market-based conservation and sustainable development, and an early attempt to intertwine market-based conservation with economic and social development.

The story below is about the winners and losers left in the wake of drug discovery from nature through one of the largest U.S. federally funded bioprospecting programs, the International Cooperative Biodiversity Group (ICBG). The ICBG officially ran in Madagascar from 1997 to 2013, making it one of the longest-running conservation and development programs in the country's history. It specifically focuses on an ethnographic account of a group of Malagasy scientists on an international "bioprospecting hunt" in one of the most remote and ecologically unique forests in northern Madagascar, in search for the elusive blockbuster drug. This was not just an individual scientific endeavor, but also a national project, as the sustainable stream of revenue through the commercial discovery of new drugs, energy, and chemicals could bring a much-needed source of income for biodiversity conservation.

Here I will discuss the dream of drug discovery from nature, set up as the model of the blue, green, and bio-economies, and show that sustainable development intervention has morphed into a host of other capitalist-intensive industries. I describe the skilled, unskilled, and deskilled labor force who collect this biodata, and who have set up the biodiversity offsets described in the next chapter. I conclude with a discussion the future of bioprospecting through biomimicry and extremophiles in the deepest, darkest corners of nature, from deep-sea trenches to outer space.[16] The chapter demonstrates that plant collections over the past twenty-five years (and even earlier during the colonial period as shown in the previous chapter) have helped amass power to those select individuals, institutions, organizations, and private sector firms who now broker environmental policy in Madagascar.

Bioprospecting as Conservation and Development

The late twentieth century was a breakthrough moment for bioprospecting. Advances in drug-related therapy and access to user-friendly drug screening technologies, such as computerized databases and robotics, increased the demand for a unique array of biogenetic resources used for new commercial natural products.[17]

While the majority of this technology was designed for pharmaceutical firms, emerging biotechnology, agriculture, and seed companies saw the potential in genetic resources to revolutionize the life science market.[18] Consumer trends were driving markets toward natural products and, by the end of the century, the bioeconomy was estimated to be worth hundreds of billions of dollars, expanding into the horticulture, cosmetic and personal care, fragrance and flavors, botanicals, and food and beverage industries.[19]

For a while, demand for the biogenetic resources came from large-scale public laboratories, but as discovery of new drugs and other products became more commercially attractive, private sector firms increased their involvement. In search for the next blockbuster, large pharmaceutical firms expanded their portfolio of drug discovery methods, including their natural product divisions.[20]

The discovery of natural products from genetic resources also coincided with a growing anxiety for the environment, including mass species extinction and, in particular, deforestation and climate change.[21] The majority of the most biologically desirable flora and fauna were sited in tropical and subtropical ecosystems of the Global South. Researchers argued that the vast numbers of species would provide the diverse array of samples needed to find novel bioactivity. The more species tested against a drug target, the greater chances of finding a "hit" or novel bioactivity demonstrating effectiveness against a disease such as cancer or diabetes. Biodiversity in tropical ecosystems means that distinctive and novel chemicals are built up over time as defenses against the vast array of predators. Indigenous and local people have known for centuries of the efficacy of some of these chemicals and have a long history of using nature for medicinal purposes. Scientists felt that if they could somehow tap into this knowledge, it could help lead to possible new discoveries and save millions of lives while also saving them millions of dollars and countless hours of research.[22] If this knowledge were to lead to a novel discovery, significant returns could be then be made through some sort of monetary compensation.

American anthropologist Cori Hayden states that bioprospecting was a way in which the pharmaceutical industry could engage through "pegging the 'value' of nature into quantifiable measures of industrial worth."[23] As she notes, "One of the key aspects of this articulation of biodiversity as a field of potential loss is the formulation of a storehouse of information not yet catalogued and thus with a value that can only be imagined." Underlining the rhetorical power of the capture and control of biodata is the prize of a conservation opportunity through the market of finding a lucrative drug. It is, as Hayden puts it, "the potential for transforming plants into information, and information into a patentable product, that allows proponents to label bioprospecting a form of sustainable—ecological friendly—economic development."[24] And while they began as small-scale integrated conservation and development projects (ICDPs) around buffer zones of national parks, market-based programs such as bioprospecting became, for many conservationists, the *only* way to save the last remaining hotspots on earth.[25]

Yet most of this biodiversity and knowledge was, of course, at quite a distance from the major drug development centers in the United States, Europe, and Japan. This geographical imbalance has raised a number of issues concerning the proprietary use of natural resources and the knowledge systems associated with their commercialization. Addressing a worldwide concern over the potential misuse of biological material and the fair return of benefits back to countries and local communities, the 1992 Earth Summit in Rio provided the first global regulatory consortium dealing with genetic resources. The summit created a number of access and benefit-sharing protocols, conventions of intellectual property rights, and frameworks for regional biodiversity agreements, which culminated with the signing of the Convention on Biological Diversity (CBD).[26] The CBD was an important milestone in bioprospecting, paving the way for large-scale collaborative international drug discovery programs, with industrial pharmaceutical firms joining forces with host-country laboratories.[27] These new bioprospecting schemes were founded on the logic that discoveries would be monetarily rewarded in a predetermined compensation deal, or an access and benefit-sharing agreement (ABS). This led to the subsequent Nagoya Protocol access and benefit-sharing agreements.[28] For many involved in the projects, these schemes amounted to the proverbial "win-win" scenario, providing the motivation to finance conservation efforts in tropical ecosystems, where both biodiversity and traditional knowledge of medicinal usage is deemed to be of the highest value.

Nevertheless, far from settling concerns, the ratification of the CBD sparked a highly polarized debate, coalescing around the proprietary use and control of biogenetics, and fair compensation for those who supplied the resources and intellectual property. Advocates put forward images of research scientists in the medical and pharmaceutical industries working side-by-side with local shamans in the search for medicinal remedies. Critics, however, forecast "piracy" of biological material and knowledge, expressing concerns about the "commercialization and potential monopolization of a fundamental constitutive of life—genetic material."[29] These concerns mainly stem from a long history of colonial extraction of natural resources and commercial exploitation of vulnerable populations in Third World collection areas. Many within the industry have also complained that the CBD simply adds yet another layer of bureaucracy to the already-complicated and expensive process of drug discovery.[30]

The architects of these agreements promoted bioprospecting as setting off a "new age" of natural product commercialization, incorporating "transparent" and "ethical" collection practices, including the return of benefits and technology transfer to countries that supplied the resources. It was, for many, the model conservation and development program, which provided incentives for locals to protect their biodiversity and addressed conservation funding and income generation.

Two examples of early bioprospecting programs designed on this "sustainable development" model are the INBio bioprospecting project, signed in September 1991, which brought together the U.S. pharmaceutical giant Merck and Co. and the National Biodiversity Institute of Costa Rica on a joint research-sharing platform, and the project for this chapter, the U.S. federally funded International Cooperative Biodiversity Groups (ICBG).

According to the terms of the INBio agreement, Merck paid $1.135 million for a two-year research and sampling project, and royalties on products subsequently commercialized from plant, insect, and other biological samples. In return, INBio was to contribute 10 percent of the budget and 50 percent of any royalties to biodiversity conservation efforts in Costa Rica. The conservation and development community heralded the INBio agreement (1992–97) as hitting the core tenets of the CBD.[31] Moreover, it signified a major transformation in the way public-private collaborations between the "developed" and "developing" countries operated, paving the way for subsequent bioprospecting initiatives.[32] The ICBG, funded by the National Institutes of Health, expanded into one of the most ambitious bioprospecting projects ever attempted by the U.S. government, involving eight collaborative research groups who conducted research in multiple countries at any given time.[33]

The INBio and ICBG projects were designed to address many of the issues in the Earth Summit and the ABS agreements of the CBD and Nagoya. These issues included the safeguarding of intellectual property for those engaged in research, the conservation of biodiversity, the promotion of economic and social development in the Global South and, most importantly, the equitable distribution of benefits from the exchange of biodiversity and appropriation of ethnobotanical knowledge. To the U.S. scientific community at the time, these bioprospecting projects allotted the opportunity for global recognition, financing, and natural resource sourcing on a scale previously unforeseen. Bioprospecting, which was the shining star of the Earth Summit, had come of age.

On the Bioprospecting Trail

In November 2005, I was fortunate to participate in a bioprospecting expedition. It was the height of the collecting season for the ICBG. I was the lone *vazáha* (foreigner, in Malagasy),[34] in a group of Malagasy scientists, researchers, guides, and porters.[35] Equipped with our headlights, plant clippers, sisal-sacs and *antsy-be* (elongated machetes), we passed through the heart of Madagascar's "vanilla triangle" and into the remote Daraina region and a relatively unknown forest, nestled within the isolated Loky and Manambato River valleys in Madagascar's northernmost province, Antsiranana.

During our expedition, I listened attentively to the lead botanist of the group, Jean, as he explained the purpose of a Global Positioning System (GPS) device to a group of porters he had hired from a nearby village.[36] Jean remarked: "The device was given to me by a U.S. botanical repository. It is used to locate my exact position when I collect a plant, and when the plant is analyzed in a laboratory in the U.S. and found to have interesting medicinal qualities, I then can return to the spot and collect more." After hearing Jean's description, I was interested to find out what exactly the porters knew about bioprospecting, so I chimed in to ask what they thought of people from the United States being so interested in plants growing in their backyard. One porter responded in the Antankarana dialect: "What does a *vazáha* want with plants? Sapphire, gold, yes, but plants?" The fact that this was the first time the porters had seen a GPS device was not very surprising since this remote area welcomes relatively few outsiders, but I was interested to learn that the porters had not heard of the team's reason for prospecting for plants. Surely, they would at least be informed of the purpose of the trip? Were these hired laborers not *part* of the bioprospecting mission?

The research team (including myself) was perceived by one French-speaking porter as "vazáha qui suivent le chemin des anciens prospecteurs," or "whites who follow the path of previous prospectors." Those hunting for minerals and other riches have a long and varied history in Madagascar, and especially in Antsiranana. The province is home to a wealth of precious gems and historically was the site of the biggest gold deposit in the country.[37] The porter, who might have been hired to carry bags for mineral prospectors in the past, was now part of a new type of prospecting mission, one in which the correct biology and chemical tinkering might produce a new drug with a value vastly more significant than gold or sapphires. In addition, as with previous missions, the porter was attempting to benefit as much as he could; however, in this case, his pay only amounted to a one-day wage of 5,000 Malagasy Ariary (approximately $2.50). In the opinion of one leading con-servation practitioner, whose organization was part of the bioprospecting mission, "They [the porters] are happy to cash in their bioprospecting chips. It's like someone who gets paid to shovel in a gold rush." This statement adds to the multiple complexities of benefits and burdens that exist within bioprospecting, and is reminiscent of Martinez-Alier's notion of inequality in the Global South, that "the poor sell cheap."[38]

Drug discovery from nature is hard work. Not only do you need to find and collect unique samples never tested before for bioactivity; you also need to transport them to an advanced laboratory far away in a developed country. This is extremely expensive and time consuming, with only a small chance that anything collected will ever lead to a commercial product. In fact, some estimates place a one in ten thousand chance of finding any form of bioac-tivity, and even less on developing a new drug.[39]

Yet it is valuable to ask who exactly "the poor" in this story are, and what exactly they are selling cheap. Over the years, critical scholars have writ-ten quite extensively about bioprospecting and the effects on communities who have engaged in the market-based conservation practice. Yet biopros-pecting is quite complex, and the issues surrounding the practice cannot be deduced as simply "biopiracy."[40] Nevertheless, critical activists and scholars

Figure 7 Locally hired porter looking on as botanist collects herbarium specimens for bioprospecting. (Photo by author)

have been very influential in galvanizing grassroots support and awareness of the potential dangers to indigenous property rights that can occur under mass bioprospecting programs.[41]

In fact, many Malagasy are quite positive about the practice, especially those scientists who have benefited from doing research funded by foreign institutions. Some locals even say that if conservation were to take hold in their communities that they would be quite happy, as it would allow them some protection against "outsiders" coming in and exploiting their forests.[42]

Hayden, working on medicinal plant markets in Mexico, says that new species discovery can also lead to a reevaluation of nature, unleashing a global conservation effort that can protect these treasures for years to come. The conservation logic behind this "taxonomic call to arms" is to create a biodiversity inventory rapidly and efficiently, before a crisis of massive deforestation and species extinction wipes out the world's biodiversity.[43] Following suit, the capitalist logic holds that any species not chemically screened is a lost opportunity to find a potential blockbuster drug.

Medical geographer Bronwyn Parry carried out an instrumental investigation of the "fate of collected material." In order to overcome the difficulties of botanical collecting in difficult physical settings, archives of digitized databases of biogenetic material and associated botanical information have been built worldwide. These plant repositories are fully stocked so that scientists and pharmaceutical companies have ready access to the botanical information and material they contain and can carry out research without recourse to the countries of origin.[44] Schroeder notes that it is this increasing trend toward what Parry calls "re-mining" that has made accountability "up and down the production chain next to impossible." Scientific and technological change in bioprospecting parallels a shift in the labor force away from local traditional healers, since complex and often difficult social relations between scientists and healers have historically posed barriers to traditional collection.[45]

On one level, I hope to expand on this important work, providing clarity around how far the industry has extended beyond this early research. I provide voices of Malagasy scientists as they navigate the contemporary world of drug discovery from their island's nature, presenting these as windows into the complexity of creating value from nature through science. On the one hand, a lot has changed in terms of bioprospecting research and, on the other, in Madagascar things have changed very little. Early on in the history of bioprospecting, the scientific community saw bioprospecting as providing opportunities to finance plant collection at a previously unimagined scale—the perfect melding of conservation and capitalism.[46]

Nature's Historical Role in Drug Discovery

The use of biogenetic resources for medicinal purposes is well-documented in historical texts.[47] The *Atharvarvada*, a Hindu text from the Indian subcontinent dating back to approximately 1000 BCE, has countless references to the use and preparation of medicinal plants for healing and spiritual purposes;[48] the landmark medicinal text *Chinese Materia Medica* (125 BCE) describes thousands of discoveries collected over various ancient ruling dynasties;[49] and Arab civilizations were at the center of medicinal plant use and knowledge from around the fifth to the twelfth century. Physicians of this period published some of the most influential medical practices using herbal remedies known at that time.[50]

In the colonial period, European expeditions set out across the globe to collect economically useful flora and fauna, yielding a bounty of useful medicinal and industrial plants. Collections from these botanical missions added prestige to European botanical centers and repositories, such as London's Kew, the Hortus Botanicus in Leiden, and the Parc de Jardin du Roi in Paris.[51] People's local knowledge of the plants' use, brought back alongside the collections, helped scientists to subsequently isolate valuable medicinal treatments. Some sources claim that the discovery of a medicinal remedy for intermittent sickness originated with Jesuit priests who observed the indigenous people of Quito, Peru, using cinchona bark (*Cinchona officinalis*) in a decoction to reduce shivering and cold spells. This medicinal remedy subsequently led scientists to isolate quinine to treat malaria.[52] Another important discovery brought over from the New World to Europe was a treatment for amoebic dysentery derived from ipecacuanha root (*Cephaelis ipecacuanha*), which is still used to this day as an emetic for respiratory infections. Other botanical gems, like teak, rubber, and sugar, were also taken to European collections in order to breed germ-stock in situ. During the collection periods, colonial institutes, including the French L'Office de la Recherche Scientifique et Technique Outre-Mer (ORSTOM) and the British Tropical Development and Research Institute (TDRI), set up experimental gardens with newly acquired exotic and introduced species.

The rapid advancement of organic chemistry in the nineteenth century led chemists in Europe to some of the earliest remedies rooted in mineral salts and natural-based metals. These discoveries, spurred by growth in the scientific fields of organic and advanced analytic chemistry, paved the way for the discoveries of ergotamine (1818), quinine (1819), atropine (1831), tubocurarine (1835), and cocaine (1860).[53]

The first National Cancer Institute (NCI) plant-screening program (1955–82) included 14,000 crude natural products (plant, marine, and microorganisms) sourced from 60 different countries.[54] In 1986, spurred by the discoveries of the anticancer drug paclitaxel (Taxol) from the bark of the Pacific yew (*Taxus brevifolia*), the NCI natural products program began a second phase of natural products research. In Phase II (1986–97), NCI signed the first of its five-year multirenewable contracts with three different major botanical institutions: the Missouri Botanical Garden (MBG), the New York Botanical Gardens (NYBG), and the University of Illinois.[55] Scouring the globe, collaborations netted up to 60,000 plants, microbes, and marine organisms, and the NCI eventually tested 500,000 extracts for antitumor activity. The advances mentioned above in drug discovery science and genomics in the 1990s, coupled with the advent of new high-throughput screening, automation, and information technology techniques, paved the way for researchers to run thousands of biological extracts at rates commercially attractive to large-scale private laboratories and pharmaceutical firms.

The Rolling Back of Nature

The dream of using nature as a starting point to discover drugs in the private sector seemed to fizzle quite rapidly. As with many market-based conservation and development programs, just as bioprospecting programs were getting started in the early 2000s, big pharmaceutical companies' attention began to wane, and they began to seek out new "more scientific" and "rational" ways to find the next elusive blockbuster. Approaches utilizing chemical or synthetic alternatives were gaining steam,[56] and the advent of new combinatory-chemistry, or "combichem," approaches fostered a "chemical revolution," which quickly gained mass appeal in the pharmaceutical industry. Combichem is a process that uses robotically created combinations of synthesized molecules to derive many "virtual libraries" of new chemical structures and enables derived molecules to be "tailored" to fit the desired molecular target in disease-related therapy. This breakthrough in technology promised to shorten the time and lessen the financial burden in bringing home the "blockbuster" drug. For many in the life-sciences industry, combichem's promise of quick and inexpensive drug discovery was very appealing and resulted in a shift away from natural products, which many in the industry saw as "clumsy" and hard to work with because of their natural complexity.[57]

To write off bioprospecting would be to misrepresent the practice. Private biotechnology firms and laboratories continue to collect natural products and have diversified their approach. Beyond just researching drug discovery, bioprospectors are putting their energy and resources into an array of natural products, including industrial biofuels, agrochemicals, functional foods, cosmeceuticals, and nutraceuticals. There is also a shift to collect extremophiles: organisms that thrive in nontraditional sites, including unexplored extreme biophysical or geochemical environments, such as deep-ocean thermal vents or alkaline and saline pools.

There have also been many other changes taking place in the industry. No longer are companies scouring forests for new whole organisms; nor are they collecting traditional knowledge as a primary lead for new medicines. It is the information harvested at the genomic level that has become the real prize. There are very few large projects where whole species organisms are collected. Rather, as researcher Sarah Laid argues, the focus of collections has shifted away from the "exotic" and far-flung locations of the past and over to "a gaze downward and inward towards the ordinary."[58] In fact, soil microbes found locally have become the new bioprospecting hotspot. One can now just collect microorganisms from marine water and soil in local sites, where the majority of microorganisms remain undescribed.

The increasing use of microorganisms is a real game changer for bioprospecting. No longer do scientists need to go to tropical rainforests to find something unknown; they can just head to their backyard and easily locate microbial diversity. As quoted by Laird, "It would take us lifetimes to sort through what we can get our hands on from this region, so there is no need to collect overseas."[59]

Over the last ten to fifteen years, the scientific community has come to realize that the real value in organisms is the genes that enable organisms to make the compounds that they do. Bioprospecting in the 1990s emphasized the whole organism, but now the emphasis is on the genes. We need to incorporate this into our models for benefit-sharing.

Genetic resources these days are mined from already-existing collections housed in digitally shared online libraries. By far, this is the biggest shift in how drug discovery involving nature works. Digital collections of genomes, mainly involving microorganisms, present many new challenges around the governance and understanding the provenance of material.

This new bioprospecting of microorganisms is generally left to smaller biotechnology startups, working closely with larger companies who can take forward the development of the discovery and marketing of the drugs. In effect, companies have decentralized a part of their research and development activities. They can provide a biotech start-up with money and equipment, while the cutting-edge technology and research comes from the start-up in return. Academia and government laboratories still conduct drug discovery work, but most of the large public labs still conduct outdated automated biochemical screening, whereas more flexible smaller companies conduct genomics-driven drug discovery.[60]

The Natural and Social Barriers to Drug Discovery

Still, for many bioprospectors, the most prized sites for finding bioactive natural products are in the Third World, quite a distance from where the research and development take place. Therefore, bioprospectors have faced considerable transport obstacles while trying to source the biogenetic resources. First, in order to conduct adequate natural product research and then develop a drug, a large amount of raw material is sometimes needed. For example, for the discovery of Eli Lilly's "wonder drug," Oncovin (vincristine), from the medicinal plant rosy periwinkle, up to one ton of the raw leaf material is needed to extract just a single ounce of the active alkaloid for the usable drug. While original source material has been reported to have come from Jamaica, it was then commercialized by sourcing tons of the raw material from Madagascar or India (two sites chosen by Eli Lilly for collection).[61] However, this long and complex supply chain became a financial liability for Lilly, which eventually led the company to experiment with the production of periwinkle on plantations in the U.S.[62]

Tropical forests are also difficult to transverse. The locations, usually far from population centers, can take weeks to find, and in the rainy season roads are routinely impassable. Unpredictable climate conditions are also problematic: violent storms, cyclones, and heat are only a few of the climatic obstacles that bioprospectors encounter. Furthermore, once material is located, it may also be very difficult to get hold of. Flowering samples, often located high in trees, and buried roots may pose significant obstacles to obtaining enough source material to conduct even the most basic of screenings.

Even after promising resources are located and collected, the challenges continue for bioprospectors. Plant material must be dried quickly, since it is vulnerable to fungal contamination and rot. Furthermore, once in the laboratory, many tropical plant extracts are found to contain high amounts of phenolic compounds, with some plants said to have up to 90 percent of their weight in tannins, making them unusable for bioprospectors. The structural complexity of natural products can slow down the ability to trace bioactivity, making further fractionation and identification of the compound difficult.[63]

Finally, replication can be extremely problematic, and after countless hours of research and resources spent, the "novel" bioactive molecular structure may actually be a duplicate discovery.

Alongside these issues, there are a number of social and regulatory obstacles that bioprospectors need to traverse when trying to access nature for drug discovery. For example, the Convention on Biological Diversity (CBD) is currently one of the premier documents in environmental governance. As discussed above, the key innovation of the CBD was to establish a framework for the development of national strategies to negotiate access to biogenetic resources in return for adequate benefit-sharing. However, its vague language left signatory parties with considerable confusion as to how to move forward and address the concerns of intellectual property rights and the distribution of benefits from commercialization.

One of the most widely cited protocols of the CBD is Article 15(1), which describes the individual nation-state's rights over its natural resources. Bioprospectors maintain that, rather than facilitating a country's resources for access, some states have imposed a draconian interpretation of the Article, making it harder for foreign researchers, and even host-country scientists, to access source materials.[64] Second, in many source countries, national access and benefit-sharing agreements are not uniformly consistent with the Article, leaving bioprospecting programs to basically "write their own rules" to access the resources, sharing of benefits, and prior informed consent.[65] As an industry representative from Dow AgroSciences noted:

> A big obstacle has actually been the biodiversity treaty [CBD], which is not standardized in developing countries. Local people have different ideas of what companies are going and not going to do, and it is just too expensive for companies to do one-on-one negotiation with everyone involved.[66]

Many scientists note that inconsistent bioprospecting rules and regulations have slowed down the process, and that the misinterpretation of the CBD by many national governments, coupled with unrealistic expectations of benefits, has led to a difficult environment in which to operate effective bioprospecting programs. And even though one of the main promises of the CBD is to provide unencumbered access to a country's biodiversity, for many in the industry, it considerably slowed down the process of discovering drugs from natural resources.

A second regulatory obstacle for bioprospectors chiefly concerns the protection and enforcement of intellectual property. Internationally, the World Trade Organization's (WTO) Agreement on Trade Related Aspects of Intellectual Property Rights (TRIPS) was the main force behind much of the debate on intellectual property rights (IPRs). TRIPS put more emphasis on private property rights, especially in respect of intellectual property, providing an easy way for companies and individuals to patent discoveries made from nature, based on scientific or traditional knowledge.[67]

Critics claim that the patenting of biological life under biotechnology and bioprospecting breached ethical boundaries, thus setting a damaging precedent for corporate control of life-forms. Also, many concerned scholars and activists remarked that "traditional" knowledge was never formally accounted for under the agreement, and rather than offer protection, TRIPS made it easier to privatize knowledge under the framework of capitalism and patent rights.[68]

In the end, the difficulties in identifying all parties and ensuring that they were informed and would share in the benefits that derived from commercialization ultimately caused confusion and misunderstandings in the industry about the correct way to move forward with the collection of natural products and the ethical return of benefits. These problems ultimately led many in the pharmaceutical industry to explore other options to discover drugs. One such option included computer-generated and synthetic-based compounds. These new efforts were seen as "rational" and "scientific," and promised to bring new drug discoveries without all the political entanglements that came with having to collect specimens of nature.

Bioprospecting a Biodiversity Hotspot

For three decades, Madagascar has held a special place on the bioprospecting map. As one of the world's "hottest" biodiversity hotspots, scientists believed the extremely high flora and faunal endemism contained unique potential for drug discovery. The International Cooperative Biodiversity Groups (ICBG) in Madagascar is composed of U.S. and Malagasy private and public international organizations, research institutions, and companies working together in a large-scale collaborative effort to discover novel pharmaceutical and industrial products. The two leading private U.S. research institutions, the Missouri Botanical Gardens (MBG) and Virginia Tech, share the role in scientific decision-making in the ICBG. MBG's primary role is to conduct inventories of botanical material, whereas Virginia Tech administers biochemical drug screening in the United States. Conservation International (CI), the largest partner in terms of overall budget finances and infrastructure in Madagascar, administers the rural conservation and development projects (known as microprojects) associated with the project. CI then subcontracts the maintenance of the field-based microprojects to a regional Malagasy NGO called Service d'Appui à la Gestion de l'Environnement (SAGE). The two corporate partners in the ICBG, Dow AgroSciences, (a subsidiary of the Dow Chemical Company, based in Indianapolis) and Eisai Research Institute (a U.S. pharmaceutical subsidiary of the Tokyo-based Eisai Co.), conduct natural product tests for novel industrial, agricultural, and pharmaceutical products, respectively. And lastly, the two national research centers, which act as partner institutions within Madagascar, include the National Center for Applied Pharmacological Research (Centre National d'Applications et des Recherches Pharmaceutiques—CNARP) and the Center for Research on the Environment (Centre National de Recherches sur l'Environnement—CNRE). Both Malagasy research institutions conduct limited drug discovery testing; however, CNARP's main role in the ICBG is to facilitate collection permits and prepare the chemical and biological extracts for export to the United States, while CNRE helps in the collection and production of marine extracts.

ICBG-Madagascar is a bioprospecting project that originated from the Biodiversity Utilization in Suriname Project (1993–97), led by Dr. David Kingston from Virginia Tech, a public land grant research university with its main campus in Blacksburg, Virginia. This first stage of the project in Madagascar, which was designated Phase I (1997–2003), was designed around collections surrounding Zahamena National Park, in the eastern forests of the Toamasina Province. Phase II (2003–8) led to the ICBG-Madagascar team being granted another five-year round of funding, and Phase III (2008–13) expanded the project to include terrestrial and marine locations in the northern province of Antsiranana (previously known as Diego-Suarez).[69]

While the project in Suriname used a mixed approach, drawing on both ethnobotany and "random collecting" of plants, the project in Madagascar only used random collection. Ethnobotany utilizes "traditional" knowledge gathered from local shamans and healers to guide the selection of species to collect and test. Random collection, designed to collect species in mass quantities, makes use of an arbitrary selection of plants found in a designated geographic range. This switch allowed the project to collect many more herbarium specimens and plant samples to run against disease targets. Random collection was particularly important because it allowed researchers to sweep up thousands of samples at a time. This benefited those agencies and institutions looking to categorize and collect as many species as possible and turn them into the bio-data now used in further market programs, such as biodiversity offsetting (see chapter 4). A commercial bioprospecting scientist told me, "*We'll do the best that we can, collecting whatever we can. For us, the more extracts the better; it all depends upon the number we can get our hands on*."[70] In the end, as we see below, this vacuum-cleaner approach was a major advantage for MBG, the lead botanical agency, whose dual mandate was the identification and systematic collection of new species in Madagascar. They gathered thousands of new plant species to add to their collections. It also allowed the project to circumvent the sticky issues of contracts with local healers, who provided the medicinal knowledge of plants in the area. This is particularly problematic in Madagascar, where this knowledge is not necessarily believed to be owned by one person or village, but rather passed down from a healer's ancestors.

This is not to say that those interacting in the ICBG did not benefit from the program. Australian geographer Daniel Robinson wrote about the advantages of the ICBG program to communities and the scientific networks developed during the fifteen years of bioprospecting samples.[71] And while some observations by researchers, both in and outside the project, have certainly demonstrated the short-term economic and social benefits of ICBG's commitment to drug discovery and capacity-building in the country,[72] I demonstrate below how structural unevenness, which has become the conundrum for conservation and development, persists to this day. Such uneven development has been the biggest problem facing the industry, both within scientific laboratories participating in the program and those conducting bioprospecting among and between communities.

The Malagasy Scientists and the Conundrum of International Collaborations

During fieldwork, I interviewed Lala, a leading Malagasy pharmacologist at her university laboratory in Antananarivo.[73] We sat in Lala's office discussing her drug discovery research from freshly collected plant samples. A few minutes into our chat, the sound of shattering glass emanated from her lab. Lala looked up at me in shock and ran over to see what had happened. It turned out that one of her students had dropped a large beaker. Upon her return, I asked if everything was okay, and she responded:

> Yes, everyone is safe, but when we lose just one piece of glassware, it is a big loss. . . . It takes months to get anything from abroad, and they are too expensive for us but we cannot do work without them.[74]

Scientists frequently described their laboratories lacking some of the most basic scientific equipment. In another interview, Robert, a lead chemist at the National Pharmacological Laboratory of CNARP, told me that he had just returned from a six-week trip to Virginia Tech in the United States (part of the ICBG), where he followed up on some interesting leads found in Malagasy plants. Robert noted that his trip was very productive:

> The work I conducted in those six weeks saved me up to one year of research time back in Madagascar. It is very frustrating to return, because I have to wait for everything to be set up and ready for work. I have to wait for chemicals, organic solvents to be shipped in. At Virginia Tech, the organic solvents flow like water from the tap.[75]

Back in his lab in Madagascar, Robert now lacks the essential equipment to efficiently identify chemical compounds and access ready-made organic solvents, and his role as a lead chemist therefore diminishes to that of a research technician. To conduct experiments, researchers must correctly identify bioactivity with vital identification equipment and information technology, most of which are not available in Madagascar. These materials include large amounts of organic solvents to make extracts, high-tech equipment, high-throughput screens required to identify bioactivity, and spectrometers to separate molecular compounds. The Malagasy chemist's role is simply to facilitate access and extraction by their counterparts in the United States, and it is the latter who conducts drug discovery in lieu of direct participation. A few years later, I interviewed Robert again, in order to ask how things were going since we last spoke. He told me that things have only worsened since the five-year economic crisis, which officially ended in

2013. The economy has never fully recovered, and national research laboratories still lack some of the most basic materials to conduct bioprospecting research.

Yet, even to this day, I am still finding scientists who say that if anyone has benefited from bioprospecting, it is the researchers directly involved in the project. Due to work carried out by the ICBG, over a hundred peer-reviewed journal papers and book chapters, conference papers, and proceedings' publications have come out of the cross-collaboration. The MBG alone has over 125 staff members, many of whom are Malagasy scientists and researchers trained through the organization.

Much of this published research highlights Malagasy scientists who might not otherwise have had such global recognition. For example, one article focuses on an important antimalarial, antiplasmodial project that came out of the ICBG partnership,[76] and one day I asked a Malagasy ethnobotanist what he felt about professional benefits from the ICBG. At the time, this professional had over twenty years' experience working on both systematic botanical collections and drug discovery surveys with local healers (*Ombasi* in Malagasy):

> For the U.S. researchers, they'll be known after discovery. But for us, for me, I need money, and also I'd like to be well-known. . . . So yes, I need publications for whoever finds the plants interesting.[77]

Some scientists were adamant that the international collaboration brought some wins. One substantial "benefit" that ICBG members highlighted is the biological screening laboratory centered at the Department of Pharmacology at CNARP. With funding from the ICBG and partnerships made with other Malagasy and foreign institutions outside the ICBG,[78] CNARP has been able to set up an advanced screening laboratory for bioactivity, using extracts found in both plant and marine organisms. The laboratory screens mainly for malaria, using techniques and materials from the ICBG. As noted by Eric, one of the malaria scientists working at CNARP:

> You know I think with only the ICBG project . . . I am now at an advanced level of research on malaria. If I was outside [the project], I am sure that without ICBG, I wouldn't have been here; I mean in the discovery process

of new molecules on malaria. For the time being, I think we are the only one in this part of Africa.[79]

CNARP seeks external subsidies, and collaborates with foreign companies and laboratories. Their position places the research center in a particularly difficult position to negotiate, and an unequal position of power. This adds new credence to a comment one CNARP researcher raised, "*Either you collaborate or die.*"[80]

This frustration was expressed to me by the lead administrator on the ICBG in Madagascar:

> To find a drug is utopist; rather we need new policies; for more of a push for new molecules, we can't, as Malagasy partners, force them to find new drugs and for us to receive the new benefits calculated in terms of new molecules, not new drugs![81]

For most of the Malagasy scientists, the inability to control the subsequent steps in the process of drug discovery (since their scientific knowledge and material gets exported to the United States) takes them out of the process, and with few major breakthroughs since the beginning of the project, this adds to the feeling of mistrust and possible theft:

> Collaboration is not a problem, developing the contract, which enhances more capacity building. You can control a little bit more . . . to just separate compounds in crude extract and send them out of the country, then you are negotiating in a "blind position." I think that the CNARP researcher was not really involved in the scientific aspect of the research within ICBG. We have undertaken most of our activity on the development of collecting plants and preparing extract, but we have not really participated in the scientific research. We have not enough equipment and materials. We need a tight cooperation with researchers here, we have to go to the U.S. and undertake there a part of their work, especially on chemistry, on the chemical level.[82]

A leading chemist working for CNRE, the other main Malagasy ICBG lab, commented, concerning the connections between identification of molecules and securing research funding:

For the time being, our laboratory is not able to identify molecules, so we need to send the extract to VIPSU, and it's them who does the identification. It has always been like that, but it might be changed in the future, not just for our laboratories, but also for those of the university or private laboratories. But this is the only way to get funding, and also to reinforce our capacities to do research.[83]

The frustration posed by not having the correct equipment to check for bioactivity in-country was also expressed to me by a higher-level ministry official:

The different types of collaboration between institutions always, it is an official license . . . to come into Madagascar and pick up the crude material, you pick up the plant and make the extract and this is the end of the process. No sharing of benefits, just a license for the foreign companies to come and pick up material. If Malagasy can at least check or screen for bioactivity here in Madagascar, then they can negotiate in a way that makes things even.[84]

Things are a bit different at private laboratories that have relationships with private pharmaceutical firms. CVO, which was developed at the semi-private laboratory IMRA, or the Institute de Pasture, is fully private and, while it has some access to equipment, it lacks the skilled training to really "play the game" of drug discovery. The chemists have a very special role in drug discovery, because they hold the "end product" of the process. For example, after the active molecule is isolated by the chemists, this end product then becomes a very marketable item; yet they are at a different level scientifically, described by the lead scientist of one of the Malagasy institutions as a

marche de dupe, a fool's bargain, or rather, it is not even a bargain, it is already a rigged market. The transfer of technology is never a fair exchange with such a poor country. The bigger never wants the smaller guy to grow. What they want is the crude material from the third world. Transfer the technology, never. We are furnishing the developed world with raw material; they change it and resell it back to us in a drug.[85]

This internal tension plays out in different ways within the ICBG. For instance, collected plant material is kept as secret codes within the department of botany. This, I was told, was done to *"keep the name of the plant secret until recollection can take place,"*[86] and in a sense, keep vital information of plant identification from members of the ICBG who might wish to take some important leads and go forward on their own independent research. The scientist above added:

> Is it a way to compete with the "big guys," if they are ready to give one million dollars to a developing country for drug discovery, to build a lab? No, what I see is our plant knowledge–based discovery as a funnel . . . a way to channel the little advantages we can give in a very direct way.[87]

In contrast, scientists in the United States, Europe, and Asia have witnessed a research revolution in the life sciences and engineering. Advances in biotechnology and nanotechnology, including digital information systems, generally referred to as bioinformatics, and also robotics and artificial intelligence (AI), have opened new and exciting pathways for life-changing innovations in medicine, food, and the materials science industries.

Specifically, innovations from genome mapping and gene editing, known as CRISPR (clustered regularly interspaced short palindromic repeats), and synthetic biology (syn-bio), provide endless possibilities, and potential pitfalls, in the creation of tailor-made products from nature. Yet Malagasy science, like many industries in the Global South, has been left out of this scientific revolution. Lacking in resources and equipment, many are unable to conduct even basic scientific research. In addition, while the idea of discovering drugs from their plants remains a source of hopeful optimism, when observed critically it is also the only type of research that many can still do. Gene editing and combinatory chemistry is just not an option, so rudimentary bioprospecting their own plants becomes an only hope, and the final line of defense during a crisis such as COVID-19. A leading scientist at a well-funded laboratory responded to the question of whether they felt supported by the state:

> Yes and No! Yes, insofar as the research center is financed by the government through state subsidies, quite simply allows the research for staff to be paid and water and electricity to flow; and no, one just needs to look at the percentage of funding granted by the state for research and you see none of this happens.[88]

Malagasy scientists are at a breaking point. Funding for scientific research has never been a top priority. Even before the pandemic, the Malagasy were feeling marginalized, and since then, this has only exacerbated their position in the global scientific order. Even though President Rajoelina has defended Malagasy science and the work carried out by the country's scientists, many top Malagasy scientists feel they have not been supported by the state. This makes sense when one examines Madagascar's investment in research and development (R&D), which is quite dismal. The latest figures, collected in 2018, show that investment into R&D was at 0.015 percent of overall GDP, a significant drop since its high in 2003, when it was still low at 0.338 percent of GDP.[89] At the time, a leading Malagasy scientist explained the situation as follows:

> We [Malagasy institutions] are a passport for plant material. The ICBG is not a bad program; they provide some materials. But it holds back when it comes to the very vital research interests that we need. So it is just not the best for us Malagasy.[90]

I received a similar reaction from an independent Malagasy scientist at the Institute de Pasteur Research in Antananarivo:

> What will Madagascar as a country gain from this? We will continue to just be a plant provider.

This feeling of disenfranchisement was echoed later by another research scientist:

> When it comes down to it, I dream of discovering new drugs because this is what I was trained in. I am just happy to contribute to scientific knowledge because fifty years down the line you never know. . . . But when I think of the ICBG project, I feel cheated as a Malagasy, as a scientist.[91]

The scientist's tension here is driven by what he sees as an unfair and forced collaboration with the ICBG. The scientist, and the Malagasy research institute in which he works, must collaborate in order to get research funding to conduct the most basic of bioprospecting experiments. This "collaboration conundrum" has existed in Madagascar since the late 1990s when the Malagasy government's research agencies were burdened with considerable financial hardships and national research institutions (i.e., CNARP and CNRE) had to look toward multilateral and bilateral donor institutions, such as the World Bank and USAID, for research funding. To overcome any shortfalls in funding, research institutions are now pressured by the Malagasy state and development agencies to produce and sell commodities and provide associated services that can be leveraged for research funds. Under this policy of market-led funding, Malagasy research scientists are led into research collaborations that provide bioprospecting services to cover even the most basic materials they need to conduct their own drug-discovery experiments.

The Local Bioprospecting Eco-precariat

Although the more technical jobs are rotated among the scientists in the group, the hard, manual bioprospecting work requiring brute strength is reserved for the hired laborers. For example, considerable effort is required to dig out rocks and soil debris to get to plants' roots. These day laborers are invaluable assets because they are paid for their knowledge of the collection sites, their ability to quickly locate flowering plants and trees, and in particular their willingness to take on the heavy physical labor tasks for little pay.

During research, I found that most bioprospecting workers were quite happy to accept a small daily wage for their manual work. Yet beyond this onetime wage of 5,000 Ar. (roughly $2.40), their involvement in the project was minimal. In a survey conducted in the three village collection sites, only half of the rural residents interviewed (including the day laborers) held any knowledge of the collection team's activities, and most were unfamiliar with the organizations that were working on the project (e.g., the Missouri Botanical Garden). Of those rural Malagasy who had heard of the researchers, less than half knew that they were collecting plants to make drugs or medicines; moreover, very few had any other detailed knowledge of the bioprospecting project.

One of the goals written into the programmatic structure of the ICBG is to support economic development and conservation interventions in rural areas. These conservation and development programs offer economic incentives to the Malagasy government and regional *communes* to conduct rural-level microdevelopment projects or microprojects. The ultimate purpose of

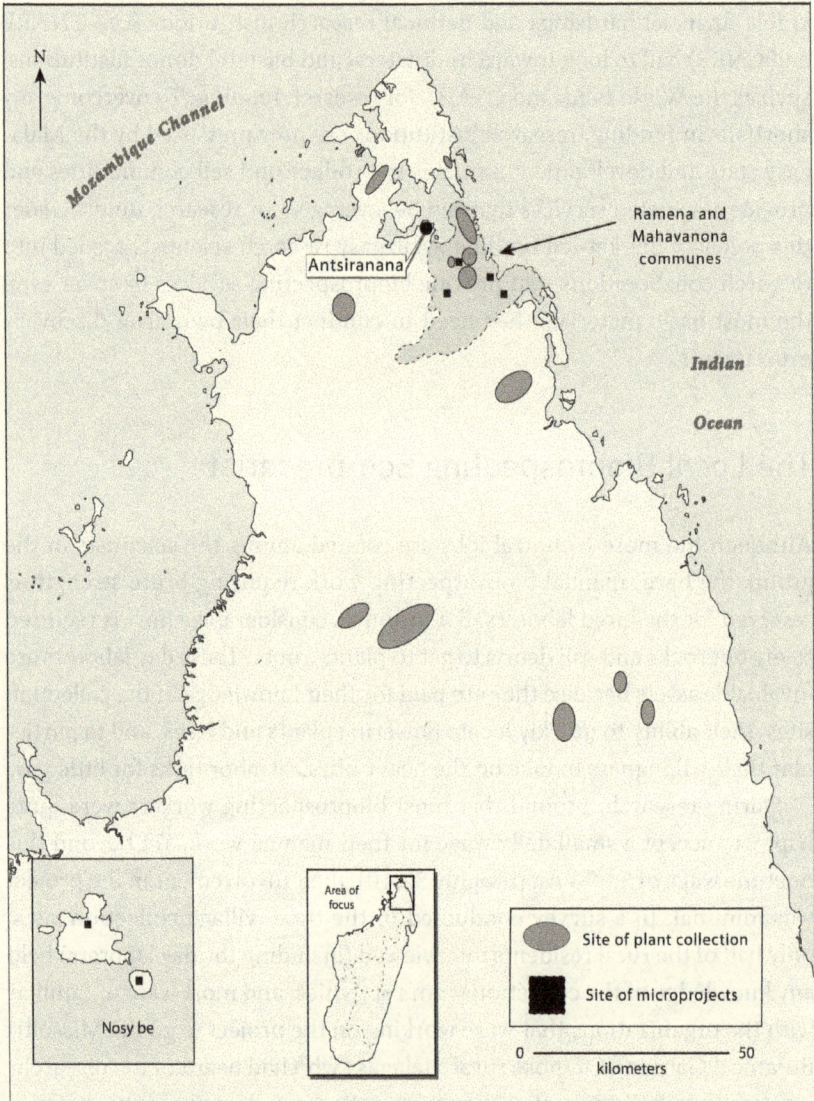

Figure 8 Map of bioprospecting and proposed conservation sites in collaboration with the ICBG project. (Author WFH)

the microproject is to provide tangible "compensation" to "local communi-ties" for the collection of the biological material.[92]

The term "upfront compensation" was used by many of the informants and participants to describe payments given to the rural inhabitants for their participation in conservation activities in areas located near sites of collection.[93] The funds for the project are provided before or during collection, and are upfront or prior to any other monetary returns such as royalties or milestone payments that may be received after any discoveries are made. The logic behind the compensation scheme is rather straightforward: drug discovery is a complex process that takes a great deal of time (estimates to bring a drug to market are upward of ten to fifteen years). This upfront payment accordingly provides an example of benefits that may be gained from protecting biodiversity. Moreover, the project holds that by providing rural Malagasy with economic alternatives through income-generating activities, they will begin to reduce charcoal production, pasture burning, and other "unsustainable" livelihood practices, and begin buying into long-term conservation stewardship.

In Malagasy, the use of the term *misy valeur be ao* (there are treasures there) in this context is particularly significant, because it reflects knowledge of the researchers' mission to extract resources that may be both "unique" and "quite valuable." It also indicates that rural residents are quite aware that their forests are among the richest biodiversity hotspots in Madagascar if not the world, and it is important to control access so that rural residents can begin to benefit from anything extracted.[94] As Mamy, a rural resident in the village reflected:

> The microproject is not compensation given by the researchers for collecting plants; *it is to get the people out of the forest.* We haven't seen the compensation yet [from the researchers] . . . and it will probably never come. The important people will keep it. That's why I said that it is better that *we* take over the management of the forest.[95]

Like other peasant economies, rural Malagasy depend on the forest for a number of livelihood resources. Unlike the southern and eastern regions of Madagascar, where forests are used for *tavy* or upland shifting cultivation agriculture, forests in the north provide multiple economic and social benefits, including the provision of timber for construction, fodder for livestock, fuelwood, charcoal, medicinal plants, and fibers.[96] Furthermore, for many rural Malagasy, forests are particularly important social meeting spaces, and many places hold sacred cultural significance.

For many scientists involved in the project, the possibility of benefits being returned to the villages, either in the form of labor payments or the microprojects themselves, is viewed favorably. For example, the virtues of the microprojects are mentioned repeatedly by ICBG representatives as "a method of giving something back to the source country and especially the 'local community.'"[97] The permanent representative of the MBG, for example, commented on the completion of a Phase I microproject:

> There was the construction of a bridge and granary. I was there during the inauguration [of the bridge]. They were happy to see their work accomplished. The real advantage of the bridge allowed the villagers to get to the hospital easier.[98]

But how do the rural inhabitants view the microprojects? For some, such as the president of one of the target villages, the microprojects seem like an equivalent exchange for their country's resources:

> I think it is equal. They came here only once. They spent one week and gave us 14,000,000 Ar [roughly $6,900]. So, I think it is equal. Maybe they got more compared to what they took but whatever we get is already fine for us.[99]

More generally, within the three villages surveyed, residents' accounts of the microprojects were mixed, and participation in the microproject and implementation was largely limited to a few individuals in each. And even though all three villages had microprojects that were actively being, or had previously been, constructed, most residents had little or no knowledge that they were even occurring, let alone that they were linked to the ICBG.

As noted earlier, beginning in the 1990s, the Malagasy government developed significant aid relationships with powerful environmental institutions and organizations; the latter provided much-needed foreign aid based on environmental conservation conditionalities or targets met by the government. The proliferation of environmental NGOs strengthened the authority of botanical agencies to influence environmental decision-making, such as drawing boundaries of protected areas, and thus facilitated select access to bioprospecting samples. As noted by a highly trained Malagasy botanist working for the ICBG:

> I collect for inventory purposes, just to know what was out there. The purpose of our trip is to do collection, and there are no exceptions for the woody plants, we take everything with a fruit and flower.[100]

It is estimated that in fifteen plus years of the ICBG collection, approximately 8,000 plant samples and 1,000 microbial samples were used in the analysis for drug discovery.[101] However, these totals pale in comparison to the number of herbarium samples that MBG was able to amass during their fifteen years of working with the ICBG. Rough estimates of 203,984 herbarium specimens are housed in digital catalogs. These collections are a significant source of pride for the teams and are symbolic of the dual mandate of the project—to discover natural products and seek out new species for identification. MBG has worked in Madagascar for over twenty-five years. Fourteen of those years were under the remit of the ICBG project. MBG's Herbarium Africa collections total approximately 650,000, with roughly 150,000 added each year.[102]

The Elusive Link Between Drug Discovery and Conservation

While one might say that the Malagasy have benefited in many ways by bioprospecting, the goal of conservation has received mixed reviews at best. Certainly, some small groups of local Malagasy who have participated in bioprospecting trips, or communities that live near collection sites, have received some small compensation. Another major benefit afforded to these projects has been the ability to identify new sites for conservation based on collection. However, the elusive and, some would say, the most important outcome of linking drug discovery and conservation for locals living closest to these sites is still quite far away. Contrasting views were expressed to me on this—the first by the director of the ICBG:

> Yes, I feel that we [the ICBG-Madagascar bioprospecting project] have done good work, even though at this point we do not have a drug to show for it. Probably the biggest benefit to date has been the establishment of the Montagne des Français as a protected area. There have also been numerous training activities and infrastructure improvements . . . in addition to the economic development projects funded by the upfront contribution.[103]

Village-level benefits usually come in the form of small-scale infrastructure or social and environmental development benefits. Like the other projects in the book, these projects are developed through community consultation with village-level associations. While the ICBG did this somewhat independently, the other two chapters discuss projects built on the Community Based Associations (COBAs) that existed at that time. The community agreed to provide up to 10 percent of the project costs, usually paid out through in-kind labor.[104]

In the first phases of the ICBG near Zahamena, the communities chose to use what they were given in "upfront" compensation for rice silos and to rebuild schools. Another major intervention was an access bridge for the community and tourists to the area.

There were a number of microprojects reported by the ICBG that were welcomed by the villages, including a boat and a motor, fishermen's storage, a cooperative and meeting site, life vests and nets, rebuilt bridges and schools, market shops and new gates, vegetable gardens, and money for the procurement of seedlings and equipment for propagating native species over eleven different project sites. In Phase II of the project, which took place in communities in the north and northwest of the country, schools, public meeting halls, and water wells were selected as means of compensation. The ICBG built troughs for animals and four wells for villages in the area—water was raised as a major priority for the area, and so the community chose to allocate funds primarily toward this. Some money had been allocated to a project relating to the rearing of chickens; however, several stakeholders had pointed out that this was not a very successful project because a disease infected the chickens and they all eventually died. In the third phase, the ICBG funded the construction and rehabilitation of schools in the area, toilets and sanitation, and a common house for meetings and events. In 2010, they also purchased boats and fishing equipment for at least one community. In late 2013 they contributed two wooden motorboats to communities, for fishing and to enable them to bring products to the town. In addition, they purchased life jackets, GPS, binoculars, and outboard motors. These boats will also be used for ecological monitoring and patrolling the marine-protected areas to monitor prohibited fishing activities.

However, not everyone agreed with how benefits were exchanged, as observed by Henri, a rural resident:

> If they [ICBG] tell us that they get new drugs from the plants, and not hide it, maybe there will be a benefit for people in the village. Still, we didn't know why they had gone into the forest, and it was only after they came back that we found out. In the end, we didn't know if they had their collecting permits or not.[105]

There were also multiple perspectives as to whether the compensation was tied to any conservation. As expressed by two rural residents in Sabatinava:

Andre: I think it is a good project because of the common benefit. If they will get new drugs from what they have found in the forest, everyone in Madagascar will all benefit from the drugs. And we expect a lot in return.[106]

Lano: We haven't received any benefit. They just collected the plants, put them in a big bag and they were gone. Nothing! However, they said that one day, they may be able to make something [drugs] from the plants and that can be our benefit. At least, that's what they said.[107] Only the president received benefit from these researchers because he went with them. He has also taken some people from the village with him, but they are the only ones who get money. They gave him money and gifts. Moreover, he didn't report to his people what they did there. Even people in the village don't know what they are doing there.[108]

Paradoxically, many rural Malagasy in this area also expressed excitement about the new "protection" their forests were slated to receive. They were under the impression that the new conservation programs would remove restrictive centralized control and place the forests under a management regime in which they themselves would be able to regulate resource access and use on their own terms. However, these expectations seemed to reflect a skewed interpretation of what protected status really meant. When asked to describe "protected areas," locals suggested that their purpose was to set into place "some sort of control mechanism to exclude outsiders" from extracting valuable local resources.

Conclusion

Early bioprospecting in Madagascar contributed threefold to the evolution of conservation and development, and the rollout of the modern bioeconomy. First, it was one of the first fact-finding mechanisms for benefit-sharing in institutional research and small-scale programs at the local level; second, it developed a professional labor force, both in laboratories and plant-collecting sites; and third, its knowledge base was significant in terms of how the labor force might administer programs moving forward. As we will see in the next chapter, the development of biodiversity offsetting, or the measuring and calculating "like for like" forest ecosystems damaged by mining, requires years of botanical and zoological knowledge about the unique flora and fauna it is meant to replace.

Under the ICBG, the spaces where bioprospecting collection occurs are objectified, traditional knowledge is labeled inefficient, and labor processes are technical and industrialized. These industrialization efforts reveal themselves in a number of ways.

First, the ICBG has conducted a wholesale change in its collection methods, from those which are ethnobotanically guided, to random collecting. This change reflects a shift toward the rational and "scientific" approaches of collecting material. ICBG biogenetic resources were collected in bulk quantities without the use of traditional knowledge and thus skirted the regulations governing benefit-sharing with local healers. Scientists and researchers were able to circumvent many of the barriers of knowledge collection by basically rendering traditional knowledge obsolete.[109] The "shaman" is replaced by an industrial process of bulk sample collected by a pool of unskilled laborers. The changes in practice resulted in a process that, more or less, runs mechanically, minimizing the role that Malagasy play in the process.

Second, these concerted efforts to develop a mechanized workforce of modern bioprospectors have left the Malagasy workers with outdated equipment and meager resources which have essentially de-skilled the Malagasy scientists, placing them on the level of manual laborers. These scientists, who were at one time part of a national initiative to develop drugs from Malagasy nature and traditional knowledge, now perform the most basic bioprospecting tasks to ensure the supply of biogenetic extracts for export.

Third, to overcome the challenges of botanical collecting in difficult physical settings, there have been many advances to build archives of digitized databases of biogenetic material and associated botanical information, worldwide. These "global warehouses" give scientists and pharmaceutical companies ready access to the botanical information and material they contain, shifting the balance of power and botanical sovereignty away from the country and locality in which material was collected.

And, finally, rural livelihood spaces have been targeted for extraction-oriented conservation projects, such as bioprospecting, to collect biological resources where the diversity of plant and marine species is highest, unique, and understudied.[110] And at sites where the ICBG operates, rural Malagasy have been happy to participate in the projects and generate benefits— indeed, some were excited that their resources might soon be protected under new conservation interventions associated with bioprospecting. Yet, once again, many did not know of, or understand fully, either the conservation projects or the potential burdens that might soon materialize in terms of restricted access to resources necessary for their livelihoods. This collective ignorance indicates a low level of Malagasy decision-making in bioprospecting activities and participation in the practice in Madagascar overall.

A critical view of this collaboration might characterize the Malagasy research institutions as providing their services and access to their biodiversity to foreign laboratories in lieu of conducting their own drug discovery research. This collaboration has resulted in some strong critiques of researchers in the ICBG and closely affiliated with it, on the part of some Malagasy scientists and administrators. These critiques have charged ICBG collaborations with selling off Malagasy resources for no significant return. They have questioned whether Malagasy research institutions and agencies are effectively fulfilling their role as suitable "gatekeepers" of Malagasy natural resources and suggested that, in effect, these institutions have taken on a more subordinate role as "facilitators" of access to the country's unique flora and fauna, rather than assuming their rightful place as true partners in the bioprospecting enterprise.

Chapter 4

Green Extractives

Biodiversity Offsets and the "License to Trash"

This isn't a get-out-of-jail-free card for mines and dams in important biodiversity areas. Offsetting is really hard.

—Julia Jones, professor at Bangor University

Look, we figured they are going to get the mining contract anyway, so offsetting was our best hope. We just didn't expect the hole to be so big.

—A leading conservationist in Madagascar (Anon #4–1A)

Nickel (Ni) and cobalt (Co) are two minerals processed at Ambatovy (meaning "place of [*am*] iron-stone [*ato-vy*]"), the massive open pit mine in eastern Madagascar.[1] For those who have never laid eyes on these minerals before, they are indeed impressive. Nickel and cobalt "pillows" glimmer like silver coins as they roll out of the processing mill, ready to be shipped abroad. Currently sold in high-grade aerospace alloys and stainless-steel products, both nickel and cobalt are also sought after globally for lithium-ion batteries. These are the same batteries found in electric and hybrid vehicles,[2] and they are therefore referred to as a "green extractives"—mineral alloys critical to creating technologies that substitute for fossil fuel energy. They are considered cutting edge for transitioning societies to low-carbon economies.[3] Recently, there has been a rush on green extractives globally, including graphite, manganese, lithium, and an array of rare earth minerals.[4] Keen to get their hands on as much of this stuff as they can, corporations heralding from the world's economic superpowers, including the United States and China, are seeking sources of these minerals. Green extractives are the new frontier in a global resource grab.[5]

Impressive or not, mining for nickel and cobalt is a dirty and toxic enterprise. The Ambatovy mine is an extremely hazardous place. The mineral ore sits relatively close to the surface across a wide area, exposing the red laterite soils to erosion. The sensitive ecological systems near Ambatovy are particularly vulnerable; the mine sits on the western edge of a richly biodiverse forest, home to many species of lemur, including the iconic indri and charismatic white sifaka.

A very long (220 km) slurry pipeline precariously carries the ore across the forest to the processing and the by-product tailings facility in the port city of Tamatave. This facility is close to marine coastal habitats and exposed fishing communities, susceptible to water pollution and toxic runoff. The refining or leaching process uses sulfuric acid, a chemical that is notoriously difficult to store, and leaves the area prone to toxic contamination.[6]

Given these pressing challenges, one might expect significant resistance to the mine, especially from global environmental and human rights activists concerned for the island's exceptional biodiversity. Yet somehow in the history of Ambatovy, the very opposite has been the case. Rather than being considered an evil mining pariah, Ambatovy over the years has been welcomed with open arms by leading global conservation organizations and global finance institutions. Company representatives sit on global environmental policy advisory panels and have racked up an impressive array of awards for their work, including Nedbank Capital Sustainable Business Award's Green Star, UNAID's Good Practice Award, and the Syncrude Award for Excellence in Sustainable Development.

Ambatovy mine production, in the eyes of the sustainable business community, is seen as a "model for the green economy."[7] As one of the early sustainability managers for the mine expressed to me, "In my eight years working for Ambatovy, I may have received less than half a dozen nasty emails about our work. . . . Not too bad for the largest and longest-running extractive enterprise in Madagascar's history."[8] This lack of widespread resistance to the mine, however, begs critical questions about how the extraction of some of the most environmentally damaging minerals in the world has turned Ambatovy into a glimmering beacon of sustainability and a model for a new green economy.[9] Make no mistake, this lack of controversy is a scientific and political accomplishment.

As noted above by UK researcher Julia Jones, making biodiversity offsets is hard work. In this chapter I explore the role of the more visible (skilled) and invisible (unskilled) scientific labor who take up this task to make nature legible and measurable for net losses or gains of biodiversity under offsetting

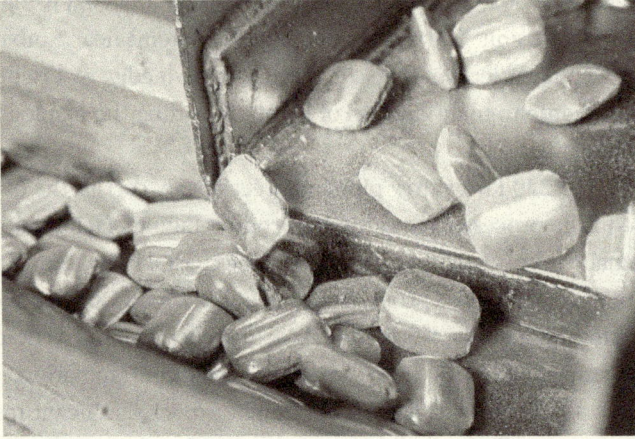

Figure 9 Nickel briquettes rolling out of a processing mill similar to those produced in Ambatovy. (Photo by Sherritt International)

Figure 10 Earth movers used to clear swaths of primary and secondary forests to build the 220 km–long pipeline in 2008. (Photo by Roberto Schmidt/AFP via Getty Images)

schemes. This is another example, not unlike the cases in chapter 3, and those in the following chapter, where a cadre of professionalized environmental-service class and unskilled laborers are mobilized and organized at the local level for market conservation programs.[10] While these projects do provide much-needed labor opportunities and income, it is vital to understand the consequences of this green economy intervention on those precarious workers most vulnerable to livelihood disruptions.

Offsetting Controversy: Science and Green Extractivism

Among all the activities that the owners at Ambatovy promote as part of the mine's special sustainability status, its biodiversity offsetting program is the shining light. Biodiversity offsetting is the practice of reducing the "net loss" of biodiversity caused by intensive mining and other extractive industries, through the protection and mitigation of intact ecosystems. Offsetting is essentially a market mechanism in which large infrastructure projects in ecologically sensitive areas can trade off unavoidable environmental damage,[11] permitting companies to wreck nature in one place by paying for its repair in another. And for some, an easy way for localized environmental externalities (e.g., deforestation) to be traded-off is sometimes far away from the site of degraded nature, essentially through a "license to trash."[12]

Yet the practice of offsetting, particularly in areas of rare and sensitive biodiversity, is extremely complex and is therefore carefully monitored by auditing institutions paid for by multinationals who are very concerned with maintaining their "green" reputation, the need for which has intensified under the new global regulatory frameworks. As a requirement for its start-up loans, Ambatovy developed its mitigation program alongside the International Finance Corporation (IFC) Performance Standard 6 (PS6), and related Equator Principles and guidance from the International Council on Mining and Minerals (ICMM). A civil society consortium under the umbrella group Business and Biodiversity Offsets Programme (BBOP) was particularly instrumental in designing the Ambatovy offsets. It was BBOP's "mitigation hierarchy framework" to avoid, minimize, restore, and offset that quickly became the offsetting industry standard.[13]

Offsetting is by no means an easy task. In fact, as this chapter demonstrates, perspectives from key architects of offsetting show just how difficult, or in some cases downright impossible, the task of replicating "one-of-a-kind forests" through offsetting schemes is.[14] These exchanges, according to anthropologist Sian Sullivan, "require the presence of measurable conservation and/or ecological restoration indicators associated with material nature, including threatened species, biodiversity, and carbon sequestered in the biomass of forests or soils."[15] New offset conservation and repaired sites are intentionally designated on biodiversity thresholds and metrics created by a new class of experts, who have collected and mined the rich biodata produced through earlier iterations of studies, starting even before the colonial period, but intensifying through the creation of a professionalized class of skilled experts described earlier as "eco-proficians."[16] It is these experts who have, over the years, grown in power and status, dictating how nature is measured, valued, and "bundled" into conservation commodities.

Setting Up the Offset: The Making of Green Sacrifice Zones

Just how did mining for nickel and cobalt become green? To answer this question, it is important to explore how biodiversity offsetting has taken off over the past thirty years. In this period, market-based solutions to global environmental crises, such as biodiversity offsetting, have been touted as our best strategy to save what is left of nature.[17] Offsetting sits alongside a host of "win-win" solutions to seemingly cascading emergencies, including what some have termed the earth's "sixth mass extinction," widespread deforestation, and disease transmission to humans from wildlife.[18]

Alongside biodiversity offsetting, interventions include the celebrated biologist E. O. Wilson's "Half-Earth," where half of the world's surface is devoted to protected areas.[19] Rewilding is another global conservation idea, which includes the reintroduction of lost species back to their natural habitats. And the most controversial, "de-extinction," mobilizes new techniques in gene-editing, such as CRISPR, to resurrect extinct species using recovered DNA. Imagine visiting your local zoo to see a dodo or woolly mammoth.[20]

These approaches indicate a deep anxiety over humanity's collective moral obligations toward the protection of biodiversity. As noted by E. O.

Wilson: "With science at its core and our transcendent moral obligation to the rest of life at its heart, the Half-Earth Project is working to conserve half the land and sea to safeguard the bulk of biodiversity, including ourselves."[21]

Cordoning off half the earth or resurrecting and reintroducing long-lost megafauna clearly represents an inflection point for global conservation. For critics, these and other market-based approaches reflect a return to "colonial-style" conservation, where some (often those in the Global South and the resources they depend on) will need to "step aside" to make room for biodiversity's greater good.[22] Yet what does it mean to sacrifice space and people for biodiversity, especially when the "goodies" or incentives for doing so, and unaccounted burdens, are unevenly shared?

For example, some have already exposed the fact that the effect of cordoning off half the earth for conservation will cause millions of the most marginal to lose access to vital resources and separate them from their ancestral lands.[23] In addition, rewilding and de-extinction have certainly had their fair share of critical responses, suggesting they can expose communities, species, and habitats to danger, often doing more harm than good.[24]

For many environmentalists and businesses, the adoption of biodiversity signifies a similar moment of crisis conservation, when populations must step aside for conservation and green growth. Sharlene Mollett and Thembela Kepe relate similar notions in their assessment of how indigenous and local communities are expected to sacrifice their rights and resources under conservation.[25] In this vein, Rosemary Collard and Jessica Dempsey refer to biodiversity offsets as "sacrifice zones," expressing that

> places and bodies that count a little bit less than others, they can be poisoned, drained or destroyed for the supposed greater good of progress. In order to have sacrifice zones you also have to have people and cultures that are available to be eroded: people and cultures that count a little less.[26]

A sacrifice zone is a geographic area that has been permanently impaired by environmental damage or economic disinvestment.[27] The term was later expanded by Lerner to show how "frontline communities, many of those populated by marginalized communities of colour and poverty, are more often exposed to chemical pollution and radiation," and are therefore regions which are deemed to be "degradable."[28] Green sacrifice zones are even more insidious when offsets are tied to extractives needed for low-energy transitions.[29] In the view of those who own the mine, we must decide where nature, and those who depend on it to survive, can die in one place, in order that it might live in another.

Mining Madagascar and the Rise of the Green Sector

Since the first Rio Earth Summit, the extractive industries have taken the lead in codeveloping sustainability platforms, mainly because the places they work in the Global South, from Bolivia to the Congo, Mozambique, and Madagascar, lack domestic social and environmental protocols. Globally, the mining sector is one of the most regulated industries, with an advanced understanding of risk. This provides both rewards and another layer of risks. However powerful green discourse is, the blowback of not doing the right thing poses significant risks. Mining companies increasingly rely on an ever-more powerful cadre of civil society groups, from environmental NGOs to scientific research institutes and think tanks, alongside the donor community, to deliver on sustainability. It is because Madagascar has such a large cadre of politically powerful environmental NGOs amenable to market initiatives that the mining industry took on offsetting with such zeal; as with many other environmental programs, Madagascar became the ideal site to pilot offsetting.

Madagascar has weak institutional governance in the mining sector, and although updated since, the mining code at the time of Ambatovy's establishment provided little guidance in the way of environmental and social protections. The one regulatory framework that does set a standard for national mining mitigation is the updated Environmental Charter (law 2015–003) and the MECIE decree (decree 99–954, amended by decree 2004–167). However, the wording for this mitigation strategy that mining firms needed to adopt was highly ambiguous at best and did not explicitly stipulate any hierarchy for avoidance, mitigation, restoration, and compensation of biodiversity; it also did not lay out much in the way of social safeguards.

This was of course problematic, since the mining sector was ready to take off with the promotion of Foreign Direct Investment (FDI) by the World Bank and IMF in Madagascar, which coincided with the expanding of protected areas during the early 2000s.[30] In fact, the 2003 Durban Vision, or the then–president Ravolomanana's plan to triple the protected areas network in the country, had significant overlap with new and existing mining claims, which needed extensive negotiations and resulted in significant tension. Because of this, larger firms, such as Ambatovy, and Rio Tinto in the south, were more than happy to enlist the support of the international community to draft plans for new biodiversity action schemes.

Yet the embrace of the mining industry by environmentalists marked a particular crossroads for the conservation and development community. Many believe that we are in a period of a climate crisis, where these minerals, essential in hybrid cars and lithium-ion batteries, are vital not just for Madagascar, but globally, for transitioning a new low-carbon green economy.[31] As expressed by a high-level mine-manager:

> Desperation is probably overstating it, but there is starting to be a sense of urgency, to ensure access to secure metal feedstock. We do not yet sell for EV batteries, but global suppliers are coming to us saying, "You know, we'd like to buy all of your production for the next five years . . ." It's because we're quite transparent, people can come and visit and see what we do and read about the biodiversity offset areas. Because of that, we find ourselves in a good position. They give us the stamp of approval.[32]

It is important not to underestimate the power the mine accrues from the overwhelming support of the global conservation community, especially given the potential economic windfall and significance of the mine to the Malagasy state. Ambatovy is the largest investment of any kind in the history of the island, somewhere in the range of $8.5 billion.[33] The mine estimates that it will place Madagascar as one of the top ten producing nations in the world, projecting 3 percent of the global supply of nickel and 12 percent of the global supply of cobalt.[34] This is quite significant, given that Madagascar is one of the world's poorest countries and currently the thirty-seventh most heavily indebted. This forces the country to welcome opportunities of Foreign Direct Investment, and Madagascar now receives 1 percent on the finished nickel and cobalt briquettes, rather than the raw laterite, accounting for roughly 32 percent of Madagascar's foreign exchange revenue. By one projection, the government is set to collect taxes, royalties, duties, and other payments of up to $50 million per year for the first ten years, and $4.5 billion over the twenty-nine-year life of the mine.[35] As one conservationist told me, "They just could not say no to that money."[36] Another expressed a very common sentiment: "If we can get them to mine better, then our job is to get in there and do just that."[37]

While this justification is not new, having it used by conservationists is somewhat jarring:

> The private sector is here to stay; they do conservation and development in some ways better than the state, and way better than NGOs.[38] In some places, they control the movement and enforcement, and therefore forests are "no go" zones, where resources are basically left alone, while other places comanaged by NGOs are left to be ravaged by villagers.[39]

As we see below, the rainforests and other village sites set aside for biodiversity offsetting are not so easily folded into the massive market transaction between the state and the mine, as formidable obstacles still hold sway. Due to this, an intensifying and expanding of commodified nature is now subsumed thorough expanded scientific labor and crystallized into biodiversity. This offsets calculative tools and complex terminology, including "leakage," "no net loss," and "mitigation hierarchies," in attempts to make biodiversity and the rural Malagasy villages that rely on it get in line and cooperate.[40]

While an imperfect market at best, carbon offsets do have agreed-upon standards by which to trade and financialize them. For the scientists and others who were charged with developing the mitigation hierarchy, biodiversity in Madagascar was not going to be so easy. As one of the leading Western scientists who studies Madagascar commented to me: "Its diversity is just too hot to handle."

> It is just very heterogeneous. If you move two-hundred kilometers to the east, to the west, you are going to have 50 to 80 percent species turnover. It has a series of very unique habitats and ecosystems that have some relationship to one another, but there is a level of heterogeneity that is extraordinary, with few parallels on the face of the earth. Hence models that you would use for exploitation of a mineral in the Congo basin or the Yangtze River basin, where you can move very large distances and have little species turnover, do not work for Madagascar.[41]

Offsetting has rarely been conducted at this scale in Madagascar, or globally in settings where the biological and geological formations are nearly impossible to successfully "reshape." As already noted, the forests of Madagascar are exceptionally biodiverse, supporting over 15,000 plant species including around 250 species of orchids, not to mention 14 different species of lemur, over 120 birds, countless reptiles and amphibians, and 25 endemic fish species.[42] To put it bluntly, if offsetting signals a shift in the way contextualized market transactions and their environmental externalities are mediated "through production of an environmental good or service removed both spatially and temporally from the 'affected' or degraded nature," then we just need to check these environments off as done for.[43] This is not to say that somewhat similar ones cannot be reconstructed, but as one Malagasy scientist who led the offsetting program told me, "by no means to the same level of diversity." This is the trade-off:

> None of these projects is perfect. You now have a multi-billion-dollar mining project in a primary forest in Madagascar in difficult circumstances. . . . It is hugely complicated because it's not only the biodiversity, but it's livelihoods and all the rest of it.[44]

The mine responded to these challenges, with a process of mobilizing scientific labor as unique as the nature itself, as Sian Sullivan notes, in order "to incorporate environmental harm into development activity and thereby turn conservation strategies into profitable enterprise."[45] In effect, Ambatovy needed to "science up," and it did so with an army of researchers ready to square the circle and manufacture the metrics as required.

"Sciencing-Up": The Making of a Biodiversity Offset in Madagascar

Ambatovy is made up of multiple mining and process facilities, and spans parts of Madagascar's vast eastern rainforest corridor. Its massive total footprint is roughly 16,000 hectares (ha), including the main mine site and buffer zones that make up 2,154 ha, and boasts 14,000 ha of biodiversity offsets.[46] The main ore site rests at the southern end of the Ankeniheny-Zahamena Rainforest Corridor (CAZ), which includes two massive, weathered lateritic nickel deposits (the "Ambatovy Deposit" and the "Analamay Deposit"), thought to be one of the largest reserves of laterite nickel in the world.[47] To the east is the Mantadia National Park and the Torotorofotsy Wetland (a Ramsar protected site). At the other end, in the port city of Tamatave, is the processing and leaching facility. This consists of multiple large industrial buildings and lies not far from the urban center of the city. Spanning the two is a massively long pipeline, which zigzags the rainforest corridor (see figure 11).

It is, therefore, no understatement to say that Ambatovy is one of the most-studied areas of Madagascar. Colonial French scientists alongside Malagasy counterparts began to identify Madagascar's mineral and biological wealth through years of geological exploration. The island's geological map was completed by the French geologist Henri Besairie in the late colonial period (1940–60), and followed up with numerous geological and geomorphological surveys.[48] The early French prospectors were particularly interested in gold and other key mineral resources, inclining tin, nickel, and copper; yet the Malagasy monarchs, leery of foreign influence, held back.[49] In fact, according to Campbell, the lack of Malagasy gold reserves led to

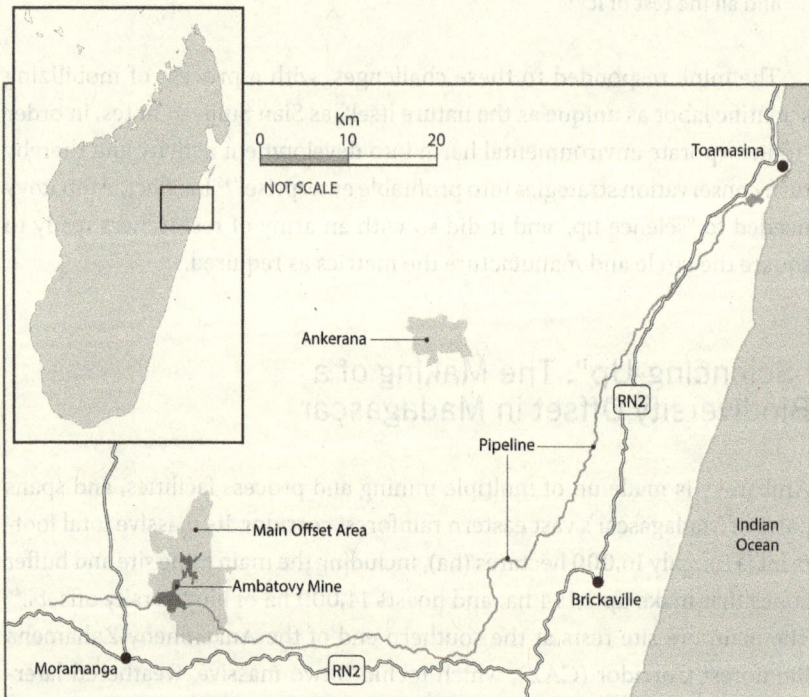

Figure 11 Map of Ambatovy mine site, pipeline, and offset areas. (Author WFH)

its eventual bankruptcy and precipitated the French takeover in 1895. And while the nickel deposits under Ambatovy were first found in 1911,[50] it was not until the 1960s that the site was seen as commercially attractive.[51] Due to the lack of capital in the newly independent Malagasy state, the large-scale commercial exploitation of Ambatovy did not happen until much later, however.

In the 1990s, the international mining company Phelps Dodge commissioned a series of environmental impact studies of the nickel deposits at Ambatovy. A second round of ecological surveys of its surrounding ecosystems, completed in 2006, included data on flora and fauna in the conservation forests immediately surrounding the mine footprint, and from other sites in the region and baseline assessments.[52] These studies formed the foundation known as the Environment and Social Impact Assessment (ESIA). This was subsequently submitted to the Madagascar National Environment Office (ONE)—the state agency that provides the permits for exploitation of the site.

Around this time, a new environment and investment law, (the decree no. 2004–167, or MECIE [Mise en Compatibilité des Investissements avec l'Environnement]) was put in place to both entice foreign investment and provide some legal transparency for investors. It came at a time when Madagascar was looking to liberalize its mining code and entice foreign investors in order to increase revenue and settle some of its foreign debt. Under MECIE, all large foreign investment projects were required to conduct an environmental impact evaluation and social development plan.[53] Given the scale of their environmental impact, this decree laid heavily on Ambatovy's planning. In order to obtain funding for these massive capital-intensive projects, the mine needed to show how it would *lessen or even improve* on the environmental damage caused by the mine; so beyond just baseline impacts, studies needed to determine the type and level of mitigation needed at the site.

A major survey that looked intensively at the diverse biological and ecological composition of flora and fauna was published in 2010 as a monograph in a special issue of the journal *Malagasy Nature*.[54] This study brought to the fore the uniqueness of the site and the extremely high biodiversity of the fauna, including several endemic mammals, amphibians, and fish. This seminal work brought together a host of well-known scientific institutions, including the University of Antananarivo and ASITY (a Malagasy bird organization partnered with the large NGO BirdLife). It also included the influential Missouri Botanical Garden (MBG), who were already identifying and collecting samples in nearby forests to use for bioprospecting (see chapter 3). These studies set the foundation for Ambatovy's first Biodiversity Management Plan (BMP) and laid the framework for the biodiversity offsetting programs to come.[55] Precise details of the project's biodiversity plan and associated mitigation programs describe a truly distinct environment around the mine that, according to some, is unmatched anywhere on earth.[56]

These studies, however, were only the beginning of what was to become a very special and close relationship between the international and Malagasy scientific community and the mine. For years, cost-benefit analyses and ecological monitoring was matched up to the baseline in order to evaluate what effect the mine would have on ecological change and local biodiversity interactions. These studies exposed the difficulties of (re)creating a similar forest to replace what the mine had destroyed. This like-for-like matching of restoring destroyed habitats is, in essence, the key to the biodiversity offsetting program and therefore central to the mine's sustainability efforts. If the mine cannot replicate what is lost, they must then come up with a different set of metrics to satisfy the banks and obtain the necessary capital needed. This is where modern finance capitalism and a professional class of green economy scientific labor came in. This crystallizing of professional science and markets was described to me this way by one of the heads of the biodiversity management team:

> When I inherited it, I changed the image to a grey-suited, serious, technical-compliance-orientated biodiversity program. So, the way I turned it around was to say, people, this is about compliance . . . You can forget about all the green NGO-type stuff. We are here to comply. And luckily, that was right, because the IFC performed standards had been applied by the bankers and we had a legal obligation to have no net loss. There was no question. Corporate CSRs are in everything, and in the end, it really became part of the corporate identity.[57]

While compliance was well known by the scientists charged with developing the metrics for the mitigation hierarchy at the time, exactly where the money was coming from and who was in charge of making the decisions over the science was less transparent. Ambatovy's loans came from fourteen different banks: five are International Finance Institutions (IFIs) and some are development banks, such as the African Development Bank. It is a very complex arrangement formed of a consortium of investors that is very ambiguous. One ecologist working there described it as follows:

> I do not think a list has ever been made public of who sits on the board of Ambatovy. It's a whole bunch of different companies and investors, many slightly hidden to some being very hidden, and hence there is no way for the world to [hold them accountable for] what they're doing, the scientific decisions that they are making.[58]

To help with this legitimacy problem, the mine brought in the "big guns": they created a Scientific Consultative Committee (SCC) that consisted of top scientists, to sort out the tricky problems around offsetting within such a sensitive ecological zone. According to the mine, the SCC helped them conform to "international standards" and avoid the difficult problems that come with such a "multi-faceted biodiversity offsets program." These scientists became, in essence, the certifiers who, through their scientific labor and the prestige of the institutions and organizations they represented, were able to legitimize the mine's activities. According to one very well-known biologist involved in the original research, the scientists gain "an important sense of egoism that their contribution or knowledge is finally recognized, and they're flattered."[59]

By no means is the process of offsetting easy, and having all these "bigheaded" scientists in one room did not always go well, as noted by one former member:

> There was so much tension, with some not feeling that their advice was taken seriously or even accepted in any detail, and some actually thought that the committee was a façade, such that the consortium was brought together to say "okay, we have this, and have the best heads in science giving their advice."[60]

In the end, many left the group, either because they were unceremoniously kicked out, or because they did not agree with the way their work was being mobilized to legitimatize the mine's actions. Along the way, Ambatovy felt that they clearly needed a bigger (scientific) boat, and they called in a highly trained transnational managerial group of offsetting expert technicians and project consultants who not only squared the offsetting circle but also redefined what squares and circles were to look like.

Science to the Rescue: BBOP's Mitigation Hierarchy and the Offset Complex

Laterite nickel and cobalt found in the tropics is surface-mined using a method of open pits, usually to a depth of a few meters into the ground. Getting to the ore is simple: the laterite is relatively soft and so there is no blasting necessary, and a well-known technique called a bench or a ten-meter strip is used, stripping the red laterite soil away and making the ore accessible. The step-like benches are crossed with "articulated haulage trucks," bringing the minerals to the ore preparation plant (OPP).

Most of Ambatovy's open pits are relatively shallow, reaching down to ten meters in total. They are, however, vast, and the huge area of exposed red soil was a shock even for a mining engineer:

> I remember flying above the mine site one day with some of the mining
> executives. You can see the forest corridor nice and green below, and the
> red lateritic thick slash, which was also the symbol of Ambatovy, by the way.
> I heard one of them say, "This is bad—you know, we don't want this. This is,
> this is really a hard site to see." Even the engineers were uncomfortable with
> cutting that much of the forest down.[61]

The development banks have to ensure that the mine is keeping up to the IFC's Performance Standards 6, and therefore, according to a mine manager, "concerns are also with the country themselves, not just the return on investment. . . . I have never experienced this kind of scrutiny ever. . . . They are much more about developing countries, industries . . . doing something that would be good for Madagascar."[62]

In order to pull this off, Ambatovy knew early on that it needed to create a new way to make offsetting work. Beyond IFC compliance, it put its name forward to become a BBOP pilot project. In a way, BBOP defined the offsetting scheme at Ambatovy, and now its mitigation hierarchy is ubiquitous globally. It created the blueprint for how it should proceed and was, by many accounts, "extremely complex and overly technical."[63] Its technical categories of organizing nature through market repair, including "mitigation hierarchy," avoidance, minimization, rescue (relocate and translocation), and offset and/or compensate, have become normalized discourse in scientific ministries, NGOs, and corporate boardrooms. This did not materialize out of nowhere, however, but was a buildup of many years of risk analysis and mitigation within the business of risk management. As noted by one of the architects of the BBOP mitigation hierarchy:

> I was working in finance in a big asset management company, and one very small part of the work I was doing was benchmarking companies on their exposure to risk because of biodiversity and the quality of their management, as a kind of benchmark framework for analysis at that risk and opportunity. Where is the last step in there, in the mitigation hierarchy, to offset the residual impacts left over, with a view to trying to achieve no net loss or if you can, to go further, preferably to achieve a biodiversity net gain? And that was just a very small part of this benchmark framework, analyzing companies. And through it, it struck me that there wasn't much consistent understanding of performance or mitigation around the world. There were no metrics for it. It was all quite subjective and not very well done. It wasn't quantified in terms of the losses and gains to biodiversity.[64]

BBOP's ability to take this discourse and transform it into language the mining company itself could mobilize in its quarterly reports, promotional materials, and green campaigns, has been phenomenal for the mine—to take one of the dirtiest industries in the world and turn it into "green extraction." A second transformative effect is the work that the mine has done to create a class of workers who can construct and develop the mitigation hierarchy in multiple settings—a veritable plug and play, through its mobilized workforce.

> Part of the challenge was that the terminology means different things to different countries. So, our idea was to put together a group where we all speak the same language . . . and could come up with standards. . . . You know, it's challenging with biodiversity. I'm very jealous of all carbon people with a nice metric. They know how to measure tons of carbon dioxide emitted and sequestered, but nobody really had the same neat metric for biodiversity loss.[65]

The language of the mitigation hierarchy plays out materially through a mixture of restoration sites, reforested and afforested areas, and newly protected areas, which were implemented at a scale never before seen by a corporate conservation program.[66] In many of these sites, the mine has pledged to provide "long-term protection through legal and managerial commitments." However, a third party manages most, if not all, of the sites, and this includes an international and/or national NGO and the state agencies, with joint funding and managerial responsibilities.[67]

The nickel and cobalt deposits of Ambatovy sat underneath a geological oddity called an "ultramafic pluton." The mine, in Ambatovy's biodiversity report, is described as follows:

> Along certain fault lines, pockets of molten rock from deep in the earth's mantle, known as "*ultramafic plutons*," were forced up to fill the cracks. Two such plutons, rich in the metals occurring deep in the earth's mantle including iron, manganese, nickel and cobalt, became the future Ambatovy and Analamay nickel deposits. Initially of a hard, dark rock, the plutons weathered and altered over millions of years of tropical sun and rain into a dark red laterite rich in nickel and cobalt, capped by a 'skin' of iron oxides, creating a hard ferruginous crust or "ferricrete" which gives Ambatovy its name (*am(b) = place of; (v)ato-vy* = iron-stone).[68]

One must not underestimate how entrenched Ambatovy geological history is with the ecological, political, and economic dynamics it has helped to shape. Social theorists Nigel Clark and Karen Yusoff argue that such "geo-social formations" of complex geology have, for far too long, been overlooked as influencers of political and social dynamics.[69] And as Nigel Clark mentioned to me, the "articulation between the strata, the hidden, mostly rocky and sometimes molten, stuff, and the earth system on the surface is a pretty lively juncture zone. That's a place where shit happens."[70]

Madagascar's eastern rainforests, first established over thirty million years ago, have since been protecting the ore from erosion. The irony is not lost on those now charged with the mine operations, whose job it is to completely upend millions of years of geological and evolutionary history. As noted by the mine, "without the forest, today's valuable nickel and cobalt deposits would have long ago been washed away into the surrounding lowlands."

Table 1 List of Ambatovy offsets and their general description

The Ankerana "offsite" offset	This primary offsetting site of roughly 5,700 ha is the azonal forest found on a similar geological outcrop to the mine, and identified by the mine, to establish "like for like . . . abiotic and biotic conditions" for threatened low- and medium-altitude forest, linked to the remaining eastern rainforest corridor.
Torotorofots, Ramsar Wetland	Adjacent to the mine, this is a conservation site that the mine supported in conjunction with Birdlife and its Malagasy NGO, Asity. It is a global site of important biodiversity. The mine provided the infrastructure and resources to establish the site.
Analamay-Mantadia Corridor	Protection and establishment of a forest corridor that links the nearby Ankeniheny-Zahamena Corridor to ensure connectivity—linking up the mine conservation sites and the protected forest.
Direct Zonal Forest Conservation	This includes conservation of "multifunctional" forests consisting of 3,600 ha of primarily zonal sites surrounding the mine. These forested areas are meant to provide "ecological services" for the displaced and existing communities living adjacent to the mine footprint.
Azonal Mine Conservation Relic	At the mine there were two patches on top of the ore body that were purposely avoided in order to keep a sample of the area that once existed and conservation of species yet discovered. This relic of azonal forest has now been destroyed by the mine footprint.
Pipeline Reforestation	The final offset includes pipeline reforestation and afforestation activities enhancing forest connectivity in targeted areas of the Ankeniheny-Zahamena Corridor.

This table was adopted from Business and Biodiversity Offsets Programme (BBOP), *Pilot Project Case Study*, 19.

The geology of the mine footprint itself was atypical, and this therefore caused the existence of the distinctive "azonal" or atypical forest, which grew over the hard iron-rich "ferricrete" crust. These forests were truly moonlike in structure and composition, characterized by short and stunted trees. They were described by a biodiversity mine consultant thus:

> This furcal crust created a very specific microhabitat at a large scale and you had this kind of bonsai-looking forest. This was something out of "Lord of the Rings." Under the influence, you know, of the east tropical heavy rainfall, you created some crazy conditions. . . . The forest was such stunted and you had interesting formations. Now we see a lot of this stuff gone today. Therefore, you were in a hotspot within a hotspot.[71]

This azonal forest meant that it was almost impossible to reconstruct the site for the offset. Given the complexity, one needs to get creative, and the scientists did just that. They found a site with the same geological formation, but far away from the mine footprint, and so it was decided that a second, similar geological oddity, called Ankerana, became the main (offsite) offset, sited over seventy kilometers from the ore body.[72]

Fixing the "Social" Puzzle

Although Ambatovy's biodiversity offsets were initially established based on years of scientific studies that constructed the detailed metrics for no net loss/gain, it did not take long for the architects of the projects to realize that finding the ecological puzzle in biodiversity offsets was the easy part, relatively speaking. Most were natural scientists developing the operational performance standards as to what (on paper) a biodiversity offset should look like. As I have already discussed, this was by no means an easy task, given the "one of a kind" biodiversity that exists in Madagascar, but defining metrics for the mine and their army of scientists was certainly achievable.

However, in Madagascar, as with other hotspots in the world, the social puzzle is a much more difficult fix. Conservationists have always had trouble dealing with the Malagasy and their interactions with nature; in fact, one might say that the inability of environment and development programs to work with those directly affected by interventions has been the major short-coming, and sometimes even an eventual failure.

Figure 12 Family members of a worker hired to lay the pipeline looking on from their upland *tavy* rice cultivation. (Photo by Roberto Schmidt / AFP via Getty Images)

This was reiterated by a senior mine manager, who declared:

It does not make much sense to base offset sites on social science metrics or the community you are trying to keep it away [from]; we were trying to be distant from people and [maintain] something that we could conserve as opposed to change the way that people behave. In the end, what we wanted was to protect an already good area.[73]

Fortunately for the mine, the areas surrounding Ambatovy are not heavily populated. In fact, the mine site sits only an hour's drive from the closest city, Moramanga, but it is estimated that there were only around thirty households working within approximately 60 ha of agricultural land within the original footprint, and a few villages were actually displaced directly by the mine site itself. The mine compensated with both monetary and in-kind payments to those displaced by the footprint.

This not to say that these populations were not affected by the mine's presence. On the contrary, some of the remote communities are very precarious indeed, living off very meager livelihoods, and any disruption to their access to patty or upland shifting agriculture for rice production could mean life or death during the bridging period between harvests, when stocks are lowest. In addition, as with most development schemes, alternative income-generating activities have met with mixed results.[74]

At an early stage, Ambatovy acquired about 100 ha in an area just west of the mine. According to the mine, affected farmers were given "a parcel of land equal to or greater than the plot they previously owned, and rice was provided during the transition period."[75] There were knock-on effects, however. For example, once an area is deemed protected, many locals will just move to another to grow rice or utilize forest resources. Once again the mine worked its way out these "social jams" by quantifying the social and political complexity of displacement and Malagasy's overall relationship with nature.[76] In particular, there was the concept of "leakage"—defined as the discounting or displacing of potential environmental damage somewhere else, such as shifting cultivators to another site because they cannot access the now-protected offset site.[77] Another BBOP term, "connectivity," or the ability to link conservation offsetting sites, is also an important feature of the social impacts of the mine. The logic being that it is easier for the mine to contain blocks of forest corridor to conservation as offsets, rather than smaller pockets (which are much harder to police), or to contain encroach-ment as Ambatovy's geographic location deep in heavily forested areas and little resources for protection makes it particularly vulnerable to shifting cultivators who use it to grow rice.

Mobilizing a Rural Labor Force

Built into the biodiversity-management scheme was the scientific labor of the communities themselves. The mine indirectly hired hundreds of rural Malagasy to monitor both the environmental changes and changes to the surrounding communities through ecological surveys of flora and fauna and through monitoring human activities such as forest burning and other potentially damaging processes. In effect, the Malagasy were paid to police themselves. Many of these rural Malagasy living in surrounding villages were not mine employees but paid as part of an existing project developed by the big conservation organizations working with the offsetting project. The mine, with the support of conservation NGOs, developed a system of Community-Based Associations, or COBAs (Communauté de Base, also known locally as Vondron'Olona Ifotony [VOI]). The Malagasy government has been pursuing a strategy of decentralization of forest resource management since 1996, under the GELOSE decree, with a scheme to devolve authority to locals.[78] These VOIs are no small thing—there are just under 1,250 VOIs in Madagascar, covering about 2.5 million ha of land,[79] about a fifth of the island's territory.[80]

In theory, under the VOI, local Malagasy set up a contractual agreement with the Ministry of the Environment and Forests and their local commune. They set out rules to manage their own forest and other resources, and these agreed-upon local-level rules are known as *dina*.[81] The logic behind the VOIs and local *dina* is that local communities' proximity to their natural resources facilitate better supervision; and moreover, since the Malagasy state has little resources or control over the island's immense rural countryside.[82] However, comanagement highlighted a number of conflicts and internal tensions with VOIs. For example, as noted in 2017 by a Malagasy conservation blogger, Nanie Ratsifandrihamanana, who interviewed those involved in VOIs, one can see "everything from internal conflicts; impotence in the face of intrusions of many types—and investors and extraction gold or sapphire miners, precious timber extraction, turtle collectors . . . but also land use conflicts with migrants."[83] Research has also found that VOIs have significant elite capture (where public resources meant for the benefit of many are channeled into the hands of a few), due to the weak support and inefficient recognition by the state of local-level agreements. The local communities are thus locally "left on their own in fulfilling their mission."[84]

Even with their drawbacks, however, these VOIs became a very helpful strategy in mobilizing the scientific labor force and management of the Ambatovy offset sites. As we see below, the mine now has access to a mobile labor force made up of hundreds of local science workers "who do the monitoring, identifying species surrounding villages, as these groups provide a real advantage in that they're on the front lines and can provide real-time surveillance on environmental change."[85] One mine administrator described the VOI's advantages to me this way:

> With the training we gave, they are able to use GPS and write correctly so that they can continue collecting data on the biodiversity offsets that you tackle, our monitoring and so on. . . . We also structured the community-based associations to run their ecological surveys as part of their commitment.[86]

In order to get an understanding of how the mine's conservation and research activities have played out with populations near the offset sites, we surveyed members and nonmembers of two locally based communities created by Ambatovy. The communities were governed by a local-level rules-based agreement that the community, mainly the association members, had negotiated with the Ministry of the Environment, based on social norms and customs, or *dina* in Malagasy.

The VOIs near the offset areas were created in 2010, and now have around fifty to sixty members, which accounts for a very small fraction of the total population in the area.[87] Around this time, the mine was beginning to expand its operations, and in the process of moving to a phase of conservation monitoring and offset implementation. The mine needed to conduct inventories to observe the effects it was having in the area and to make sure that its biodiversity plan was up to speed in terms of no net loss. The ecological monitoring was crucial to this, and the mine's work with local associations provided the right mixture of available and flexible labor to do the necessary work.

While the associations were developed to devolve resource management to local communities, the main driver was the mine's requirement for species monitoring to measure its metrics. However, although there was some training and sensibilization of the workforce, they consider their roles to be nothing more than a conservation activity. In fact, not one of them mentioned no-net-loss or annual assessment as the reasons for species monitoring, and the VOI members told me that their role was to ensure the management and protection of forests and the species inside, and in one member's words, "fight against a host of pressures, including *tavy* and logging, mainly by migrants."[88]

Many of the VOI members were contracted employees. Contracts varied from yearly through monthly to day-laborers. Many had a written contract with a daily wage of up to 10,000 Ar ($2.50) per day for day laborers, while longer-term workers could earn from 180,000 Ar ($45) to 240,000 Ar ($60) per month. The longer-term workers also said that they were able to obtain health insurance for themselves and their families.[89] And those who had contracts appreciated the work they were doing, and would agree to do more if possible. Extra perks were also described, including a hen house and chickens, and some agricultural crops, including peas, beans, rice, and seedlings of the non-native eucalyptus tree from a nursery built as part of the mine mitigation.

One of the village employees, Hari, noted that the work "was to elaborate appropriate strategy to protect the species and the forest. We had carried out patrols to reduce pressures on the forest (especially human logging) and built a plant nursery to restore the forest."[90] A few respondents stated that the forests helped "soil fertility'" and the "growing of crops,"[91] but almost none discussed that the monitoring activities had anything to do with the offsetting program. In their understanding, this was a project of direct forest conservation to combat village-level deforestation.

VOI members normally participate in ecological monitoring for the Ambatovy mine. This is usually a task that may span a year with monthly or more frequent trips into the forest, or may cover a few weeks. Members are paid directly by the mine for monthly monitoring activities for specific species of plants or animals. Here the members mentioned that they conduct inventories in transects, usually looking for lemurs, and are responsible for monitoring their habitat. Yet patrolling activities were said to be done "voluntarily." This is built into the agreement they have with the association as part of the *dina*. Many stated that they find it hard to work on the forest, especially when they are required to carry out monitoring at night. This is required when nocturnal species need to be observed, such as the elusive aye-aye lemur. In addition, the forest is located five kilometers away from their homes, and they felt these treks were a lot to do voluntarily in support of the association.

Clearly there is a significant advantage for locals to being in the VOI. Those who had some official recognition, either as a permanent or temporary member of the VOI, commented that the collaboration has enabled them to tap into the various livelihood support mechanisms (providing either rice, chickens, or agricultural seeds) and through the training in conservation activities, such as ecological inventories. In addition, many of the members of the VOI have a very positive opinion of the mine and find working with them collaborative and supportive, and one commented that they had been told "not to use force when people were captured after doing illegal activities in the forest. They always try to understand and find a solution."

One VOI member, Fifi, noted that they found the "research useful as it could help to have a deep knowledge about the forest." They are also able to access "useful trees" and look forward to working out a way to "exploit some of the forest in the offset area."[92] However, those who were not VOI members in a village did not receive such, and some of the population in the village were described to me as "people who still persist in traditional

practices and struggle in finding a job, so illegal logging practice sustains
their livelihoods."[93]

While the VOI was in consultation to develop the local *dina* to manage
the forest, it seems that those who were part of the VOI were again treated
more favorably than nonmembers. For example, the fee to use trees for construction is reduced for VOI members, and no exploitation of the forest
could take place without authorization from the association. Selection of
the association members was never a "transparent process." As noted by
one nonmember:

> The reason why they did not join the group is because the VOI members
> select people they like. There is no communication of the process. Nonmem-
> bers are the ones who are blamed when the forest is logged. They really want
> to protect the forest near to them. VOI members are not equitable with other
> people. . . . But being a member of VOI helped us to facilitate making the
> authorization to collect wood and other species from the forest.[94]

Another noted, "We are now being able to live as others. Conservation
work gives some opportunities to earn money."[95] Yet nonmembers declared
they "didn't have time to be implicated in conservation activities during the
period they have recruited," and "now have difficulty surviving, so they are
obliged to find a job."[96]

Two women members of the VOI noted that they participate in conservation activities because they are aware of the importance of the environment.
They are doing it "voluntarily and have worked with Ambatovy to create
plant nurseries and to carry out an inventory." As members of the VOI, they
have received training and participated in patrols, but said that they were not
paid for that and are doing it on a voluntary basis. They did not find it hard to
work in the forest, but said the main problem is the social issue of enforcing
conservation: "We became each other's enemies."[97]

Many of the respondents at one of the offsetting sites did what they clas-
sified as conservation work, including ecological restoration, developing and
maintaining tree nurseries, and forest patrolling/surveillance. They were also
asked to work on tree nursery management and carry out forest maintenance, including ecological restoration and tree planting. Others described
their work in more refined, research-based terms, such as animal monitoring, conducting surveys, and carrying out plant and animal species inven-

tories within the offsetting site. A few said it was more about protecting the forest, patrolling, and monitoring. Although the majority did think of the VOI as "important work," only a few mentioned that these collective benefits flowed down to each of them individually.

When asked what they thought about having foreigners interested in conserving a Malagasy forest, many had little idea, but one said that conservation is important and that if foreigners were more interested than locals in protecting their important natural resources, then the locals should be "ashamed." However, others were more skeptical of the motivation of the mine executives, saying that they were using the opportunity to illegally collect the protected species.[98] This mistrust of the mine was not uncommon, especially among nonmembers of the VOI.

Nonmembers were subject to different rules and access rights to enter the offset site. For instance, two women, Irenee and Zo, mentioned that they did not join the VOI because they do not have time, due to livelihood pressures. They recognized conservation activities were important and they agreed that foreigners are deeply engaged in protecting Madagascar's forest. One noted that at least these people were able to create jobs in the locality. They had worked for a short time with Ambatovy, exploring the forest and capturing invasive crayfish. Their short collaboration with Ambatovy, they said, did increase their income and knowledge, and afterward they felt better equipped to manage the forest. They had carried out ecological monitoring for one week, during which they received 10,000 Ar per person, in addition to the crayfish counting. Yet the amount received from conservation was very low, so they were obliged to have other full-time employment activities. They mentioned that the amount paid for extracting resources by VOI members was much less than for nonmembers, causing tension. Others noted that "yes, . . . our life has improved somewhat, as we are easily able to ask authorization to use wood," furthermore noting that conservation has "impacted our livelihoods as we stopped taking resources out; however, some people in the VOI don't stop exploiting the forest."[99]

Many of those dispossessed by the mine did receive some compensation for the land they are now excluded from. However, many claimed that the compensation was very low and insufficient. Those whose land was negatively impacted by the new road constructed by the company received no support and said that their land could not even produce half the harvest that it had in previous years.

The association requires a large amount of money from nonmembers when they wish to take something from the forest (such as *erana*, raw material for making baskets or timber). This has had a big impact, as the locals depend on natural resources to survive because "the lack of jobs in the village remains the main pressure for the forest communities."[100] Yet nonmembers have very little understanding of the local rules. Some were only hired for manual labor, to clear the area for both the mine site and the very long pipeline. Feeling a bit "left out" of continuing conservation and research work, one respondent noted, "I was hired by them to cut trees in the mining site at the beginning of the activities, but after that, the rest of the activities were destined for VOI members."[101]

Being part of the VOI is not easy, however. They are obliged to expose people and apply *dina* when illegal extraction or *tavy* occurs in the forest. Many know about the official laws from the ministry, and are aware that there are *dina* that regulate the forest's management and can be applied when there are issues in the forest (for example, illegal exploitation, logging, etc.), but there is little in-depth knowledge of the *dina* by nonmembers. A few respondents mentioned that they are able to get authorization to collect products from the usable areas of the forest and use some of the fields for agriculture. Most VOI members participated in the creation of the local rules and many knew about specific aspects of the contents. Yet there was a lot of confusion when discussing who actually is involved in their management—many still saying it was the mine or the state. Few members understood, or at least admitted, that they were themselves active managers.

People generally have some understanding of what is illegal; however, one woman noted that there are problems, "especially when the person knows that I saw him doing illegal things. They can even threaten my life," and that although "we have promised for this when we joined the VOI, it is very difficult, as it has broken relationships in the community."[102]

There were some positive responses regarding the *dina*, even from nonmembers: "It doesn't really negatively affect our living as we observed a change in the frequency of fire in our locality. I think it's a good thing, even though I am not part of the VOI."[103] Yet it does seem to come with a cost of impacting intracommunity relations and increasing social tensions. One member noted, "It improves our standard of living because of livelihood support, but we now had a difficult relationship with nonmembers of the VOI." While the majority mentioned that conservation is a good thing, having "improved their [standards of] living," this did not discount the many issues mentioned around enforcing the *dina* and the role of the association to report infractions by other family and community members.

By reporting people who broke the rules, "we have begun to be the community's enemies; as we have conflict with the *fokonolona* [village-level authorities] we are afraid wherever we go as people threaten us."[104] Another noted that conservation increased poverty because of social conflict and issues around agricultural land. Generally, when the discussion turned to loss of species, the explanation by local VOI members was that many of the native species were being lost due to "the local people," not the mine.[105]

One member recognized that, as a member, she still needed authorization to enter and obtain something from the forest but that "other community members were displeased about this, which led to conflict." "Yes, we are really restricted in our activities as we are no longer able to take wood without paying taxes and no longer able to let our oxen to the forest without being penalized by the *dina*. . . . We are restricted in our activities because of conservation."[106]

Conclusion

If anything, the Ambatovy mine is an example of the hidden labor that goes into making an offset, and therefore, making a mine certifiable and sustainable. Like other chapters in the book, this case study exemplifies how difficult it is to commodify the very rich and diverse nature of Madagascar, particularly in a place such as Ambatovy where, according to the mine's own promotional documents, it is like no other place on earth, a literal "hotspot within a hotspot."[107] In line with the related bio- and blue economies observed in chapters 3 and 5, respectively, the green economy in Madagascar has, from its inception, never been just one thing, but rather a suite of conservation and development options, many which involve multiple private and state actors and multiple scales. Biodiversity offsets encapsulate this multiplicity. Although heralded by some environmentalists as a conservation program, it is in fact more of a trade-off—an acceptance of market externalities, as one leading conservationist at WCS put it to me when describing his acceptance of the program and later acknowledgement of, and astonishment at, the effects of a mine.

However, many Malagasy entrenched in the mine are now not just passive observers but active participants in the mine's conservation activities. They count, collect, and measure species, and surveil others, some of which are close family members. They go to meetings, and they make decisions over who has access to forest resources. Although most of these activities go unnoticed and/or unaccounted for in many of the mine's quarterly reports and sustainability reports, they are essential, as this "participation" by locals provides the compliance needed to certify the biodiversity offset and make nickel and cobalt a "green extractive."

Chapter 5

Extending the Frontier

Blue Carbon and the Commodification of the Sea

> When we ran the numbers, we saw that these communities, among the poorest on earth, had found a way to double their money in a matter of months, by fishing less. . . . Imagine a savings account, from which you draw half your balance from every year and your savings keep growing. There is no investment opportunity on earth that can reliably deliver what fisheries can.
>
> —Country director, Blue Ventures

> We have remorse joining the project, because we have been trapped. Maybe conservation is good in the sea . . . but not in the mangroves. Before I thought, yes, conservation of mangroves is a good thing because we can keep them for the next generation; but now we suffer, because we have been betrayed by the state and the project. We can no longer access mangroves to cut, even to the quotas allowed.
>
> —Aina, thirty-six-year-old male fisherfolk (Anon #5–1A)

In March 2021, a special section appeared on Politico's homepage titled "The World's Hottest Commodity." The short article stated that, beyond the hype, the hottest commodity was not the wildly fluctuating and highly speculative cryptocurrency, Bitcoin, hovering at the time around $50K per coin, but rather nature, or more specifically, sectors of nature that are deemed the best to offset the world's carbon emissions.[1]

> Every day brings a new corporate pledge to fight climate change. But read the fine print and you'll see that those ambitions rely heavily on offsets, not actually cutting emissions. That means carbon sinks—soil, rocks and forests that can store gases safely away from the atmosphere—are in high demand.[2]

The article went on to say that current approaches *just* using trees would not do the trick:

> *Quality* offsets are in short supply, a problem that Exxon Mobil, FedEx and United Airlines—companies with yeti-size carbon footprints—have recently put in the spotlight. They and others are investing hundreds of millions of dollars to find more durable, scalable solutions to handle industrial emissions, because until we can learn to make concrete and fly planes without releasing greenhouse gases, *trees just won't cut it.* (emphasis added)

So, what do you do when trees cannot solve the world's carbon conundrum? You do what resource-strapped societies have always done: look to the oceans. Since its launch onto the world's stage a few years ago, carbon sequestration from marine sources and coastal mangroves (blue carbon) has become the latest of a string of silver bullets promoted as a panacea to solving the global emissions crisis.[3] It is easy to understand the hype.[4] Coastal mangrove ecosystems cover only 2 percent of the earth's surface, yet researchers have shown they have the potential to sequester over 80 percent of global carbon,[5] a rate about two to four times greater than global rates for tropical forests.[6] Furthermore, since many mangroves exist in biodiversity-rich tropics, financing blue carbon may hold the potential to deliver a windfall to poor countries and communities in the Global South.

Seeing the ocean as a panacea to the world's carbon trouble may come as a surprise to some, but many have been gearing up for this moment. Large-scale plans for blue carbon projects have been sprouting worldwide,[7] with some of the world's richest philanthropists supporting blue carbon projects, such as Jeff Bezos, whose previous gift of $100 million to the World Wildlife Fund is helping to develop blue carbon projects to "harness the power of nature to stabilize the climate crisis."[8] This idea of "industrializing" or "mobilizing the power of nature" has become a recurrent theme in new "nature-based solutions" promoted by environmentalists, as noted by a researcher at NOAA, who described blue carbon as "coastal powerhouses working for us every year and keep[ing] their stored carbon locked away."[9] However, as I describe below, it is not only a means of employing mangroves to work for us, but also the people who live near them, who now need to work for their own survival—and seemingly, more and more, for *our own* survival.

This chapter considers the expansion of blue economy growth strategies into costal marine space under the climate crisis.[10] Still taking shape, the nascent blue carbon program called Tahiry Honko (preserving mangroves), works with some of the most marginal coastal communities in southwestern Madagascar. These communities, part of the Vezo ethnic group, are highly dependent on mangrove ecosystems for fishing, fuel, and construction materials.[11] Tahiry Honko is run by an international NGO called Blue Ventures, who, alongside partners in the area, have developed very close ties with the Vezo communities.[12] One key aspect of the project is that it utilizes local community associations to administer local rules and regulations, or *dina*, over resource use, similar to the system observed in chapters 3 and 4. This, for some, is a way to get local "buy-in" and devolve resource management to these resource-dependent communities.[13] This participation is promoted by the project as a way to market blue carbon as a "high-quality" offset, due to the benefits it delivers to local communities. Yet, beyond simple participation, blue carbon in Madagascar needs the communities' skilled and unskilled scientific labor to make it work. This labor participates in monitoring stocks, surveying mangrove tree felling, and collecting data on carbon and associated fish stocks. Results show that it is a group of smallholders who are left once again pinning their hopes on a mixed bag of conservation returns and the delivery of meager benefits. This environment of increasing social tension is exacerbated by new enforcement of the state's reterritorialization, causing a revenue grab of blue carbon benefits, and the communities' scientific work effectively monitors their own demise, as their scientific training and mapping are now used against them, restricting their access to resources needed for their livelihoods.

Offsetting Our (Carbon) Sins

For many, carbon offsetting has become a potential boon to control the world's greenhouse gas emissions. The practice was first mainstreamed after a few very high-profile global policy developments, including the 1997 Kyoto Protocol (where rules were laid out under its Clean Development Mechanism) and later, the 2005 EU Emissions Trading Scheme and subsequent 2015 Paris Agreement.

Figure 13 Two local Vezo women taking GPS points to measure the extent of the mangrove forests in southwestern Madagascar. (Photo by Garth Cripps)

Carbon offsetting is a reduction-based practice that rests on the idea that industry carbon allowances can be offset by another activity that can sequester or capture that same amount of carbon from the atmosphere. Since carbon is thought to be generally the same everywhere, these "offsets" can be sited in locations far away from where the pollution was originally emitted.[14] If new, short-term emission allowances were made available in lieu of longer-term sequestering activities, this would incentivize the industry, over time, to invest in new technology to potentially bring down the total overall emissions. Since its introduction, carbon offsetting has become one of the most discussed, yet controversial, market mechanisms used to rein in the global CO_2 emissions.

In the mid-2000s, the United Nations launched a carbon sequestration program called REDD+, or reducing emissions from deforestation and forest degradation.[15] The idea behind REDD+ was to address net emissions by essentially combating tree loss[16] and engaging in better management of global forests stocks, together with incorporating wider sustainable development goals.[17] In theory, countries with large stands of trees could profit from existing forest stocks by paying people to keep them standing. REDD+

was seen as a way of mainstreaming carbon offsetting as a social investing tool by companies interested in displaying their commitment to sustainable development. It was also a tool for governments and local communities to garner climate financing for wider conservation and social development projects—particularly in the Global South, where researchers believed that the potential for carbon capture in tropical trees was much higher than in northern temperate forests.

In reality, even after thirty-plus years, some carbon markets have had a difficult time while others have just flat-out failed.[18] A recent Bloomberg article, "Wall Street's favorite climate solution is mired in disagreements," offered a frank explanation of the current state of carbon markets:

> The danger is that cheap offsets can be used to avoid the hard work of actually cutting emissions. The practice is so common that the [carbon offset] certificates are often described by critics as "papal indulgences," reminiscent of the way Catholics in the Middle Ages made payments to the Church to eliminate the stain of sinful deeds.[19]

At a meeting of the world's business elite, Mark Carney, the former governor of the Bank of England—together with Bill Winters, the Standard Chartered CEO—noted that "hundreds of executives and scientists" met to set up a global trade in carbon offsets. Asked how it was going, Carney noted, "It's messy!"

Carney's description of carbon offsets as a messy practice is spot-on. One of the main issues highlighted is that carbon projects are overly complex, formed of multinationals, state agencies, civil society organizations, financial and verifying institutions, and smallholder communities, all with multiple interests and expectations.[20] Second, carbon markets are filled with vague accounting practices and nonuniform carbon certifications, leaving room for "cheating" and overestimating the carbon sequestering potential of forests. And finally, sequestering carbon through offsetting takes place in tiny pockets of the globe, often far removed from, and out of sight of, the sites where carbon is being emitted, making it hard to regulate and keep the offsets doing what they say they are going to do.

Needless to say, global private sector investors have been generally squeamish, holding back on fully embracing carbon markets as a mechanism for climate mitigation,[21] and in practice, carbon projects are still highly dependent on funding and support from governments in the Global North. How-ever, it is not only at the funding and policy level that carbon projects have gone a bit off the rails, but also at the implementation sites. Carbon mitiga-tion has had some pernicious effects, particularly for those most socially and economically marginalized communities living closest to forests set aside for offsetting—ironically many of those same communities whose participation is central to their success or failure.[22] And yet, it is still true that carbon markets seem to maintain their momentum through the amplification of escalating environmental crisis narratives.

The "Dark Side" of Carbon Markets

For some time now, scholars have been raising the alarm regarding carbon mitigation and the broader effects on indigenous peoples and local forest communities.[23] If Global South countries were to be the main beneficiaries of offset payments, how were profits and burdens to be shared, especially with those most affected by loss of access to forests due to new carbon enclo-sures? For instance, Adeniyi Asiyanbi describes the UN REDD+ in Tanzania as another mechanism of "re-territorialisation and a roll-back of access and local control of forests which were gained over the past 25 years of con-servation."[24] Alongside others, Asiyanbi argues that seeing local forests as "carbon stocks," to be maintained and managed by the state, has serious implications for those small-scale farmers who access the same forests for their everyday livelihoods.[25] Also, American geographers Betsey Beymer-Farris and Thomas Bassett deftly draw attention to the potential for "lose-lose" scenarios of climate mitigation "that fail to integrate environmental justice concerns with conservation priorities."[26] In fact, mounting contes-tation of REDD+ processes by indigenous peoples and local communities has been observed across the Global South. Geographer Abidah Setyowati demonstrates how effects from REDD+ projects in the Aceh region have unearthed a history of prior conflicts, execrated communities, and already simmering tensions around land dispossession, as well as prolonged civil conflicts.[27] Critical social scientist Sarah Milne argues that these are some of the mismatches between conservation programs' dreams of the perfect offsetting scheme and the stark realities of uneven development; they are the consequence of socioecological "embeddedness," or the social, ecological, and cultural linkages that communities share with forests.[28] As scholars Tracy Osborne and Elizabeth Shapiro-Garza explain, "Offsets are actively produced when carbon is sequestered in trees, they have an unbreakable and continuous bond to living biomass and can therefore never be fully divorced from the place of production or the people who produce them."[29] It is these global economic values, pinned to offsets, that often clash with the local social, cultural, and political ecological meanings that many times cause projects to fail.

If anything, Madagascar is a case study of REDD+ projects gone awry.[30] Madagascar was observed early on as a leader in developing the financing infrastructure and mapping and establishing "REDD+ ready" sites for offsetting projects. However, since then, its projects have all but stalled. Beyond the discourse of local participation, local activity in project planning and implementation has been relatively poor, and even poorer in benefit-sharing in communities living at the forest frontiers.[31] The implementation has been stalled due to policy missteps and lack of meaningful legislation and roll-out. There is also growing concern as to the potential for projects to worsen poverty in vulnerable forest-edge communities, with increased calls for more stringent social safeguards and a greater participatory approach to be adopted.[32]

High-Quality Offsets? Fungible, Fictitious, and Downright Impossible to Value

In response to the challenges of REDD+ and other carbon offsetting, there have been a host of "new" and "improved" carbon mitigation programs, such as blue carbon, that seek to engage with those who have been excluded from previous schemes. One advantage advocates tout is that, unlike terrestrial forests, blue carbon is said to be of "high quality." As it was explained to me by a blue carbon offset manager, "A high-quality offset means that it comes with social and economic co-benefits to local communities, and where essentially consumers are attracted to, and hence 'more desirable' than normal offsets as they engage with local communities from the onset."[33]

However, the idea of "high-quality blue carbon offsets" or "carbon mitigation 2.0" is that it runs counter to what many experts tout as carbon's advantage as a market tool. Carbon is thought to be "fungible," and its value

therefore indistinguishable from where it originates. This interchangeability
of carbon means that, for example, one ton of carbon emitted from your
recent flight to the Spanish island of Ibiza is meant to be the equivalent to
one ton of carbon sequestered from a Mexican forest. This differs from "biodiversity offsets" observed in the previous chapter, whose value can change
from one landscape to another. According to advocates of the practice, it is
this relative ease of transferability across large distances and complex infra-
structures that makes carbon offsetting so powerful as a way for companies
to do good, as many in the private sector say, "beyond the four walls of their
office headquarters."

Yet, while carbon theoretically has definite attributes that are compara-
ble, in reality, some have argued that carbon offsets, due to their intangible,
nonmaterial nature, are not necessarily the same everywhere; therefore a
significant amount of work is required for projects to define what it is and
to legitimize it through scientific measurements, in order to make an offset
a real thing for those who want to purchase it.[34] As Lovell and colleagues
point out, carbon is not a "tangible product" but an invisible gas, which in
most large projects is hard to directly measure.[35] Furthermore, when purchased, offsets do not produce any direct service or payoff, besides possibly
the public relations benefits that firms or individuals may receive in return
for promoting their purchase. Because of this, Lovell and Liverman argue
that the discourse used to define carbon becomes extremely important in
how it is valued by voluntary offset markets:

> Certain carbon credits are attractive because they have a story associated
> with them and can be sold at a premium as "gourmet" or "boutique" carbon,
> with an emphasis on their poverty-alleviation "side benefits." In other words,
> information and knowledge about how the offset is produced—where and
> using what technology—is crucial.[36]

As anthropologist Amber Huff describes, blue carbon offsetting schemes
can only be seen as commodities in that they "inhabit a complicated, con-
tradictory, and only provisionally stabilised commodity form."[37] Huff then
goes on to say:

> Carbon offsets are disembodied, spectacular, intangible commodities. There
> is no underlying concrete asset, only an imaginary [one]. They can only be

created in the context of a project designed to materialise the unrooted and substitutable nature imagined by market environmentalism.[38]

However, one way that high-quality offset material value is made is by making the labor of measurement visible. This process of making this skilled and unskilled manual labor visible requires an army of precarious laborers who will take on the difficult work. This is unlike other cases of bioprospecting and biodiversity offsetting, where this labor is kept invisible; here the labor adds value and is explicitly exposed to bring value to the carbon offset.

This is similar to other studies, where participating labor is explicitly exposed by projects looking to promote their work. For instance, geographer Ariadne Collins speaks to the fragmented subjectivities of "indigenous and maroon community members" who get relabeled as REDD+ assistants, monitoring and evaluating REDD+ carbon schemes in Suriname and Guyana.[39] Geographers Wim Carton and Elina Andersson also speak to the subjectivities of those laboring for carbon offset producers, this time in Tanzania. Here smallholders are pressured to meet "targeted goals" of planting set by the offsetting project. These "carbon farmers," as they are called, are promoted by projects as the co-beneficiaries of carbon market schemes but, according to Carton and Andersson, find themselves in difficult contractual conditions, similar to those of more traditional crop contract farming. Their "participation" on the project is dependent on their ability to understand and meet scientifically based targets not clearly defined or laid out to them, and they are

> thereby expected to live up to a specific planting, thinning, and pruning regime, to keep track of exact land sizes and tree numbers, to combat pests and diseases, and to base their involvement in the project on a cost-benefit estimation of tree cultivation compared to a range of alternative land uses.[40]

American geographer David Lansing provides an excellent snapshot of the type of "work" done by locals to identify spaces and nature used to create value in ecosystem service projects. He describes the performance of different "collective agents" and their "calculative practices," which contribute to the combined agency of making carbon exchangeable.[41] An assemblage of scientists and local indigenous leaders "participate," through the adoption of GPS devices, in the mapping of carbon sequestering trees in Costa Rica.

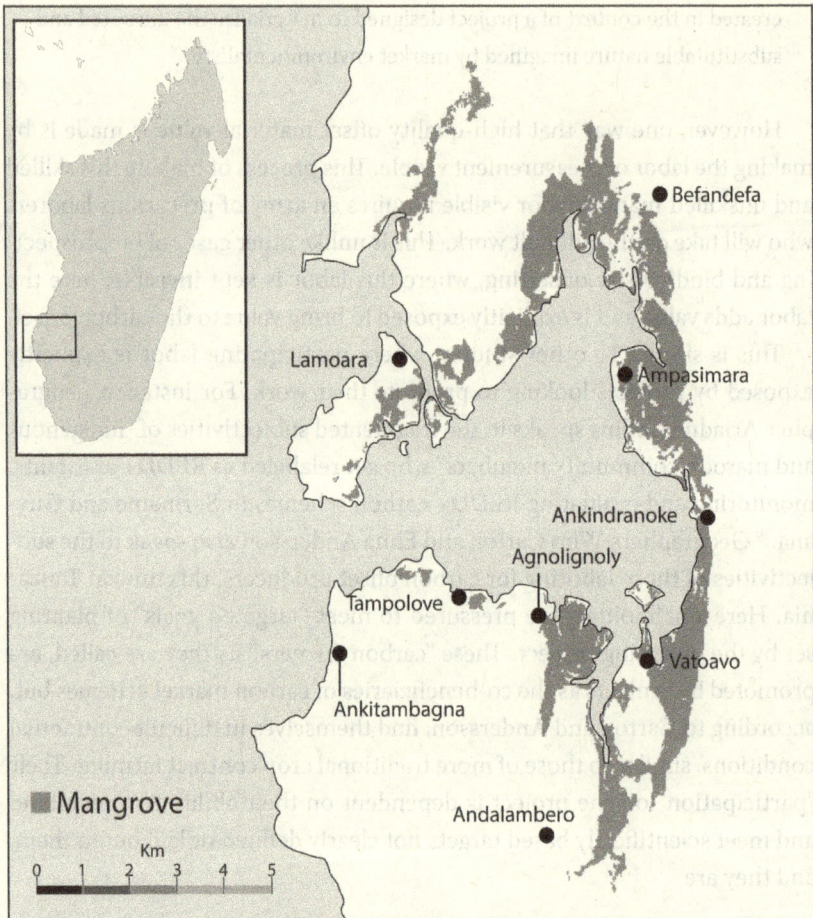

Figure 14 Map of Blue Carbon Project villages and mangroves in southwestern Madagascar. (Author WFH)

In another example, Lovell describes an increasing trend of carbon being "estimated from afar" through advanced airborne or satellite remote-sensing data, LiDAR, and algorithm-based "allometric models," which may only hasten local involvement in carbon counting work and diminish the ability to package community participation.[42]

In contrast, Madagascar has long been a laboratory to test carbon mitigation projects and a showcase for how labor "works."[43] It was at the forefront, building some of the earliest infrastructure methodologies and aligning them to local conditions.[44] In the case of blue carbon, this labor was packaged in glossy pictures, online videos, and rich descriptions of participation by its community members. This "selling of success" was one of the routes to presenting it as "high-quality."[45] However, as we see below, this labor participation came with a cost, as the revenue from bottom-up programs of mapping and managing the mangroves for carbon financing attracted attention from the state, whose interest went beyond regional conservation, and considered the revenue these new dynamic ecosystems could potentially bring to their resource-strapped coffers. This entangling of labor becomes more entrenched when speaking about "high-quality" blue carbon sequestration from mangroves in Madagascar, whose value is based on the premise that it is local scientific labor who are creating the offset through their measuring and monitoring.

Open for Business: Introducing Madagascar's Blue Carbon to the World

Africa's largest island, Madagascar, provides prime investment opportunities for those looking to capitalize from an expanding ocean frontier. Not coincidentally, the island hosted the UN-sponsored conference in 2013 on the "African Blue Economy," which led to a flurry of excitement about how the country could leverage its coastal marine and ocean territory for development. *Raconteur*, an online business marketer magazine, ranked Madagascar as the number one African nation poised for growth in the blue economy sector:

> With a 5,500-kilometre coastline, Madagascar's potential to benefit from a blue economy is huge . . . from shrimp fisheries in the West to deep-sea ports, mining and container shipping in the East and South East, the potential in the world's fourth biggest island has recently been identified by Chinese investment, which has pledged $2.7 billion to projects that range from shipyards and fisheries to aquaculture.[46]

The World Bank and African Union's 2050 joint maritime strategy document called this turn toward the oceans a "new frontier of African renaissance," promoting investments in everything from alternative tidal and wind energy to marine tourism, deep-sea mining, transnational shipping, global waste, and most relevant to this study, carbon sequestration.[47]

In November 2014, at the World Parks Congress in Sydney, the president of Madagascar, Hery Rajaonarimampianina, made a commitment to triple his country's marine protected areas in the next ten years. As one conservation NGO put it, this was "establish[ing] legal frameworks for community governance of fishing grounds while encouraging the development of economic incentive-based approaches to marine management."[48] This declaration, known as the "Promise of Sydney," turned the region's attention to the oceans as a way out of its multiple environmental and social development crises, through its two-tiered approach of local community management and market incentives/investment.[49] Nothing fit the bill better than blue carbon.

Although ideas of blue carbon had been around for some time, the formalization of ocean and marine ecosystems into carbon platforms were mainstreamed at the Twenty-First Conference of Parties (COP) Paris in 2015. It was here that nations agreed to Nationally Determined Contributions (NDCs), or the annual reduction of emissions in an attempt to limit the global temperature rise to two degrees Celsius by 2050.[50]

But unlike more state-controlled compliance regulations, blue carbon in Madagascar is a voluntary offset. It is run by nongovernmental organizations or development aid charities, which develop a "unit" of carbon dioxide equivalent (tCO_2e) or a type of reduced or sequestration credit for purchase.[51] The project must show that their units or credits of emission reduction would not have occurred otherwise, that they exist in reality, and that they can be verified. These carbon credits usually come with a certificate denoting that they are unique as they cannot be double counted, and are long-lasting or permanent.[52] These verifiable carbon certificates are produced by a host of independent verifying groups, including Social Carbon, Plan Vivo, and the Gold Standard. And unlike major nonvoluntary or compliance markets, which work with larger extractive industries and airlines, voluntary carbon markets tend to not cater to what they see as low-carbon emitting industries, such as finance institutions.[53]

However, they also tend to be cheaper and provide ease of access and a "niche" market for those more accustomed to bespoke preferences, small volumes, and immediate access.[54] They are less top-down and have fewer administrative oversights than compliance markets.[55] The variability in pricing reflects this reality, ranging from less than one dollar to up to sixty dollars.[56] While smaller in size then mostly carbon mitigation projects, such as REDD+ projects, they tend to promise more immediate impacts on the ground. In this sense, blue carbon is viewed by its advocates as a support mechanism for social and economic development for coastal communities.

This is particularly important for coastal areas in Madagascar, where communities exist without roads or market infrastructure. According to the World Bank, coastal communities are essentially a "segment of the population [that] is often among the most vulnerable and marginalized communities, without other assets such as land that could allow them to diversify their revenues."[57] Some argue that fisheries, particularly those that are artisanal and small-scale, "remain invisible," uncounted in official national statistics of GDP, yet contribute a significant proportion to huge segments of the population.[58]

At its core, blue carbon is meant to reverse this erasure of coastal communities by incentivizing marine conservation through direct payments to local communities.[59] Through their close connection with coastal communities, the project was designed to deliver socioeconomic development while building, in effect, a "conservation workforce" to enforce protection, monitor, and manage the mangrove forests through carbon financing.

While the idea of autonomous mangrove conservation seemed ideal, the political realities of blue carbon were always destined to get in the way. Very few farmers hold official titles to their land, and they have little in the way of political power. Locals were left helpless to ward off "outsiders" looking to profit from their conservation work; as we see below, this included the state reasserting its power over blue carbon. Once mangrove conservation began paying off, the state came in and securitized its assets, whereby they effectively shifted the ground from under these communities. They abruptly changed the way many of the communities, and the project itself, now sees blue carbon—and not necessarily as the panacea for mangrove conservation it was once anticipated to be.

Beyond the Hype: Blue Carbon in Madagascar

Beyond all the hype, there are only a small number of blue carbon projects actually up and running in Madagascar. Tahiry Honko is the largest and most established project, and operates in the southwest of Madagascar. The NGO running the project has been in the country since 2003, focusing mainly on marine conservation and social enterprise for coastal communities in and around the Bay of Assassins. The area comprises ten partner villages consisting of about four thousand people. Blue carbon is actually just one of the things that the organization is involved in, as the project director explained: "We are a charitable organization, and rely almost totally on donations . . . blue carbon is only one of the many activities we do, and in fact, it is really used to fund the work of locally managed conservation areas."[60] However small, it is a central feature of the overall goals of conserving mangroves, sitting at the center of the mangrove-heavy Velondriake Locally Managed Marine Area (LMMA),[61] a roughly 680 sq. km marine protected area in the south of the bay.[62]

Most of the villages working within the blue carbon project and mangrove conservation areas are Vezo, a relatively small ethnic group whose cultural identity is tied to subsistence and small-scale fishing. The Vezo are thought to be a subgroup of the much larger Sakalava.[63] Historically, Vezo are one of the most socially and economically marginal ethnic groups in the country, with a very high resource dependency.[64] These communities live in a very arid climate, not very suitable for farming, and the ocean is their main source of food. Some estimates indicate that about 80 percent of the ten villages that circle the bay rely on small-scale fisheries as their main livelihood. These take advantage of the high fishing productivity of the Mozambique Channel, known for its extensive coral reefs and seasonal upwelling.[65]

The area is densely packed with mangroves. The trees found in these diverse ecosystems are used in the area for timber construction poles, and in lime and charcoal production.[66] The area, however, also experiences seasonal surges, with "in-migration" of outside groups looking to tap into the rich resource base of the mangrove ecosystems. This has posed multiple challenges to mangroves in the region.[67] Many of the Vezo lack formal tenure rights over the coastal sites and have little recourse for maintaining or managing them. There is very little state forestry or fishery management of these vital ecosystems. Mangrove management is also particularly gendered, with a significant burden of heavy work falling to the women living in these areas, whose role is to collect fuelwood for heating and cooking. Access to the mangroves is, therefore, vital to their livelihoods.

In order to address these conservation challenges, the NGO helps to organize community associations who might provide input into the new management plans known locally as *dinas*. The NGO originally divided Velondriake

Figure 15 Two Malagasy scientists measuring carbon potential in mangrove forests in southwestern Madagascar. (Photo by Garth Cripps)

LMMA into three spatially distinct conservation zones: (1) a strict conservation zone; (2) an area of multiuse access or "sustainable" management; and (3) a zone designated for mangrove restoration.[68] Like the comanaged forests discussed in the two previous chapters, local natural resource management has been agreed upon through consultations with village leaders (*fokontany*) and association members, who settled upon the acceptable multiaccess use zones. The logic behind these conservation zones is that mangrove conservation would only be successful if these communities were allowed to access certain parts of the area for extraction.[69] The selling of carbon credits through the blue carbon program was meant primarily to be an "add-on" funding mechanism, which could also deliver social development programs within Velondriake. These development co-benefits, as they are called, are the local level, social development projects (e.g., school and community well provision) observed in the other case studies within this book, which also stem from international conservation financing. In theory, the money and social projects are meant to incentivize locals not to overuse their natural resources and "reorient" local values toward long-term environmental stewardship.[70]

At the time of writing, recent estimates suggest that at its current rate, the project produces over 1,371 tCO_2e per year, and generates about 1,300 carbon credits annually, with one carbon credit equal to one metric ton of carbon stored.[71] The project sought to deliver at least 50 percent of its profits to social enterprise projects in coastal villages for infrastructure construction, school building, and health-related initiatives. Meanwhile, 25 percent of profits were intended to go to the state as part of the larger carbon mitigation plan, with slightly less for a local marine management association covering overheads for the project.

Counting Carbon Is a Messy Business: Making Mangroves Investable Though Participatory Mapping

At the heart of Madagascar's blue carbon projects are its coastal mangroves. Madagascar has large stocks of mangrove forests—236.4k km^2 to be exact.[72] Mangroves, however, are a strange tree. Found in intertidal habitats, they can successfully grow in the most unforgiving low-oxygen, highly saline, and waterlogged settings, in which many other trees would die. Due to their unique morphology, including an elevated rooting system and an ability to efficiently filter salt, they not only survive in these harsh brackish ecosystems but thrive.[73] Interestingly, these swampy biomes have high species biodiversity, including fish, crustaceans (e.g., shrimp, crabs, oysters, and barnacles), sponges, algae, and large numbers of birds.[74]

The carbon in these diverse ocean and costal ecosystems is held, however, not in the mangroves but in the soils *under* mangroves, marshes, and seagrass, or deep in the ocean's marine sediments (made up largely of phytoplankton).[75] Blue carbon represents a huge amount of the world's stored carbon and has been forming over a long period—which is why some argue it is vital to keep it from being released.[76]

Mangrove swamps may be good at storing carbon but they are hard to navigate and work in. They have a pungent smell of hydrogen sulfide, also known by those scientists who work with it as "rotten eggs gas,"[77] and they are filled with thorns, stinging bugs, and sticky mud. One scientist comments that because of their environment, mangroves are devalued and termed as an "enemy of the state, or just a worthless swamp."[78] Yet many who work and live close to these unique environments become accustomed to, and even to some extent, appreciative of the smells and the mud. As a Malagasy blue carbon scientist said to me, "The soil smells. . . . It emanates from the slimy squelchy mud. Sometimes it is repulsive, but when you understand the mangrove soil's importance, you see it rather as a bright diamond."[79]

This excitement around mangroves is shared by many Malagasy scientists, including another blue carbon scientist, Henri:

> Blue carbon is quite difficult for those who don't know what it is exactly, because it's quite tricky. Mangroves are particular forests located between the sea and the land, and that's why they are very interesting. They store a lot of carbon in their soil and it will be kept for a millennium. It's really amazing to know that. But work is dirty—you know, working in a team in dirty slimy water. It's not comfortable. But when you are working on it and you know, it's important. You are eager to learn more. It's amazing, an amazing ecosystem![80]

Like the forests in the previous chapters, blue carbon is only salable when the mangroves' ecosystems, and the scientific work that goes behind calculating their worth, are made visible to the potential buyers. This work of making mangroves legible through measurement and calculation is mainly conducted through a process of mapping. And while mapping is a high-tech process of analysis of high altitude and satellite imagery, it is also a painstaking process of "ground-truthing," or being able to fact-check on the ground to see that the mangroves in the pictures you are looking at are actually there, not just "pixel dust." And nothing speaks to rural labor more than participatory mapping.

Local communities should be trained to gather basic data, said James Kairo, principal scientist at the Kenya Marine Fisheries Research Institute, who are involved in a similar blue carbon project: "It's very easy to build a (local) community of blue-carbon assessors," he stated. No matter how much technology there is, "we will always need the human hand." Kairo added that "success of projects depends on the local community, and among other factors, blue-carbon projects are now faster—and therefore cheaper—to develop."[81] While the rollout of these projects may be easy, it relies on the difficult work of local labor.

Mangroves seem to be one of the last ecosystems in Madagascar, and maybe the world, to be mapped. One reason for this is that they were generally left undistinguished from terrestrial forests, clustered into other coastal or littoral classes of trees. Another reason for their neglect is that they are under water half the time and thereby missed by most aerial photos or satellite imagery used to make the maps. This has changed now: the ability of new mapping techniques such as geographic information systems (GIS) and geospatial imagery allows researchers to get a more precise handle on what is out there, and therefore what is verifiable in terms of carbon capture. This includes getting precise data on subtypes of mangrove forests, as noted by Eric, a forest mapping expert who has worked for years in Madagascar's mangroves: "The three different subtypes of mangroves we observed were ecologically distinct on the ground, spectrally distinct from what we saw in our remotely sensed data, and our satellite imagery showed that they are also distinct in terms of their ability to take up carbon."[82]

This technical ability to precisely classify, calculate, and verify carbon stocks has, of course, been a boon for the blue carbon industry, and is the foundation of carbon mitigation markets. However, it has its problems, as Eric noted: "When it comes to mapping, mangroves represent their own unique set of challenges, perhaps as complicated as any forest type can be. . . . It's completely inundated with water multiple times a day, or the opposite, just sort of all this exposed mud and soil, completely changing its appearance in satellite imagery." It is these wild variations in satellite images that necessitate "ground-truthing" by locals, as Eric went on to explain:

> Through meetings, through interviews, through surveys . . . basically relying completely on community members to really understand what's happening on the ground; helping to make sense of the remote sense data and what this ground-truthing does—this is the basis of carbon stock management. But even that work is hard. It's as hard a place as you can map. And it makes ground-truthing even more important. . . . It literally laid the foundations for the project.[83]

Mapping mangroves is central to blue carbon methodology and to its very legitimacy as a product; in fact, one might say that there is no way to do blue carbon without maps.[84] Most use a combination of large open-source databases and aerial and satellite imagery, along with ground-truthing.

Unlike other blue carbon projects, this one was purposely kept as simple as possible and represented a deliberate move by the project away from complex methodologies to value and verify carbon. To ensure the carbon had high value, the project pushed for communities to be involved from the very beginning:

> Blue carbon needs to be quite simple. . . . If you look at regular carbon markets, like REDD+, it has very complicated monitoring methodologies. You need experts, but also in terms of demonstrating that the carbon generates credits. There are certain measures in place to demonstrate that to the carbon market, and eventually to buyers. And that involves third-party audits every year verifying that credits have been delivered and that trees have not been cut down. . . . We try to make that accessible and lower the barriers as much as we can by still being able to demonstrate climate benefits.[85]

The project still needed to continually monitor its mangroves and certify its carbon stocks, and then using the standard "third-party" auditor similar to large carbon projects globally, it chose a more grassroots approach called "activity-based monitoring." And so they asked select members of the village association to "ground-truth" mangroves. The logic behind this approach was simple, as highlighted by the project manager:

> And the logic behind activity-based monitoring was that, rather than imposing a very strict rule saying that the project needs to pay for remote sensing every year to make sure that the forest really wasn't degraded, or to do destructive soil sampling and things like that, we can just have the Malagasy do it.[86]

The project took two approaches to ground-truthing. One was to train locals in geospatial techniques and participatory mapping; the second was community ground-truthing. A lead project manager noted on a webinar that blue carbon "data is power," and that this data needs to be "locally owned because blue carbon is about managing resources effectively."[87] The project went to great lengths to train villagers and Malagasy scientists in mangrove mapping, and a lead technician noted,

> Malagasy and foreign people were flowing through the project, with well over twenty-five to thirty scientists coming through, and hundreds of those who were involved in some way, many who were not scientists when they began. We use things like Google Earth to show very detailed represen-

tations of where we are through the high-resolution satellite imagery, we worked with these folks to have them delineate areas of particular landings, understanding from them specifically how, why, in what ways, and where they interact with and use their surrounds. This really laid the groundwork for the establishment of proposed conservation restoration.[88]

Yet even with all these efforts, it remains too technical to do at the grassroots level, as another technician mentioned:

> Not enough communities around the world can benefit from climate finance. As it stands, developing blue carbon projects is highly technical and expensive. Unfortunately, this makes it beyond what's accessible and possible for many communities.[89]

Grassroots involvement in mapping carbon stock calculations was just one of the roles of the patrols. The second phase of monitoring the forests' carbon stocks, that of surveillance and enforcement, yielded more mixed results.

The Patrol Teams: Labor from Below

Beyond assisting in ground-truth mapping, the patrols' second mandate was to monitor the rates of logging and overextraction of other mangrove resources, such as fish, crabs, and sea cucumbers. Their task was to seek out cases where individuals were extracting resources beyond the allowance in the quota system. This was layered on top of, and overshadowed, in the eyes of the locals, the general accounting and monitoring—conducting inventories, counting trees, and measuring and mapping increases and decreases in mangrove acreage.

In fact, the patrols' tasks are multifaceted, from checking up on the sites, to making sure the signage that explains the different conservation zones is in place, to counting and monitoring any newly cut mangrove stumps.[90] Teams of up to ten villagers, both men and women, take part in patrols. Many mentioned that they do so once a month, and it can take up to a week and a half to cover the ten blue carbon sites in the Bay of Assassins. Most said they were recruited by already existing team members or though notices posted in the village. They are equipped with a GPS device, digital camera, binoculars, pens and pads, a smartphone, flashlights, and machetes.

Many felt proud to be part of the project, and to be doing the right thing by conserving the mangroves for future generations. As Hasna, a female villager noted, "It was important that villagers take charge of their resources, which is always necessary for the protection and good management of these, their own resources."[91] These patrols become the eyes and ears of the project, and thereby the main line of policing and protecting of the mangroves in the villages.

Those who are involved with the project seem to have a generally good understanding of why the project is pushing so hard for conservation of mangroves. As Lala, a female head of one of the teams, noted, "Mangroves protect the seashore and are a breeding ground for many of the fish we depend on."[92] Yet she also said that many of the villagers "are afraid of transgressing the *dina* because the *dina* are very severe."[93] Another noted that "this was meant to give us some secondary activities, for additional income, beekeeping, seaweed cultivation, sea cucumbers. We have marine protected areas in over thirty villages. If we don't reduce fishing for communities, marine resources will quickly disappear." Yet informing villagers of the rules is one thing; policing them is something else entirely.

More advanced scientific monitoring, such as soil collection and other technical measurements, are handled by smaller and more advanced scientific teams, clearly demonstrating a divide, at the village level, of skilled and unskilled workforce. These teams are made up of foreign and Malagasy scientists brought in from regional or national university laboratories. One of these Malagasy scientists told me:

> We pay them a daily wage and then we train them, but sometimes the work is just too technical. . . . They don't do it themselves. Something that is classed as manual or monitoring work, then they can do it—counting cut stumps, they can do the work, but if it is more scientific and technical, no, I need to do it *with* them. Yeah, I have to supervise them.[94]

Some of these more skilled workers in the project were already experienced in conservation work. For example, they may have already been trained in similar monitoring through the project's larger fisheries management program, counting fish species and numbers, genera of octopuses, and so on.

The Promise of Co-benefits:
Surveilling Their Own Demise

This decentralized approach has worked quite well for the NGO. They have been one of the most successful conservation organizations in recent times in Madagascar, racking up significant conservation wins and touted in the international press for their "progressive" locally owned conservation approach.[95] However, as carbon mitigation from blue carbon became more popular, it also attracted the attention of government ministries looking to cash in on what they saw as a potential revenue windfall.

After locally based agreements had been put in place and implemented, the Malagasy government put a blanket moratorium on *all* mangrove cutting in Madagascar. According to some working for the project, this was in response to the widespread mangrove deforestation taking place in the northwest of the country, but also related to the state's fear of losing out on revenue. What was reflected to me in detail by an environmental consultant and expert on marine conservation in Madagascar was that the government was, unhelpfully, restrictively monopolizing carbon credits and selling them, with the money going into the treasury coffers; thus the profits were entirely under the control and discretion of those in government, enabling them to revenue-grab in every area. It was noted that "the government does not understand the sensitivities of a market on carbon. And if they appear to be just money grabbing for the state—they're going to sell, no consequence at all."[96]

A second major issue with the project was that, early on, they decided to use the exiting REDD+ infrastructure to develop their benefit-sharing strategy. This made sense, as Madagascar has some of the most advanced programs globally. However, this centralized all carbon mitigation revenue entering the country, and put coastal mangroves in the same category as terrestrial forests. This centralizing of blue carbon control runs counter to the ethos of the project, however—that is, that blue carbon is unique and a high-quality carbon credit destined for local-level community conservation.

About twelve years ago, blue carbon scientists began to share inventories of mangrove carbon sequestration potential with the Ministry of the Environment, aiming to get mangroves included into Madagascar's REDD+ framework, as a project manager noted, who saw early on that REDD+ was a "legislative framework to latch onto, particularly in the beginning, when mangroves were not even on the map, with very little engagement or focus on them—which has all changed now!"[97] While this was definitely a way to get mangroves "on the map" and recognized as important, it also later brought with it political and regulatory baggage. It is now very hard to disentangle it from the state-mandated REDD+ framework, especially with the state's perception of a revenue windfall.

On average, blue carbon projects generate around two thousand credits—"a figure which is way below demand."[98] One of the reasons these projects become more attractive to the companies and individuals looking to buy carbon credits is that they are packed with promises of co-benefits, from the delivery of education and school-building to health care. These benefits provide the project with add-ons that are said to "satisfy the consumer and incentivize the community." But in the end, it is the NGO who is paying for these, mainly because the state has been "revenue-grabbing" or keeping the income from carbon credits at the ministry level as part of its wider REDD+ strategy. The current condition, as expressed to me by a former researcher in the project, is that

> the reason revenue hasn't flowed into the project isn't because of a lack of interest or, you know, demand from the carbon market, but it's more because of regulatory barriers put in place by the Malagasy state.[99]

These complications have also spilled over to the community members, who feel conflicted over the rollout of the project, its benefits, and the sometimes fraught relationship between the state and the NGO. The reasons for this fallout from blue carbon are many, but mainly, they are due to tying its fortunes to the only existing carbon mitigation infrastructure. The fact that communities were no longer allowed to extract mangroves, and neither would they be receiving the benefits through carbon mitigation payouts, is symbolic of the challenges of operating market-based conservation and development projects with local communities. There is therefore clearly a significant disconnect between local- and national-level political dynamics and development around carbon mitigation.[100]

It seems that while there might be economic opportunities for some, having such projects does not always bode well for community cohesion. As in the past, the influx of money and other perks can also cause jealousy and resentment. It also, however, provided a space for unexpected outcomes, such as the maintenance of benefits derived from the NGO who is now funding them, and if this includes the securitization of resources through militarization, then some see that as acceptable. As noted by one former employee:

> As the original staff of the project, we were motivated differently. We were volunteers and we worked from our hearts, with or without monetary motivation. At the moment, it is the reverse; it is personal interest that prompts others to join. The current staff give more room to the forces—that is to say, to the army, like the gendarmes.[101]

One of the main points of contention was the shift in policy by the state. The project originally divided the mangroves into three conservation zones: a strict reserve, a cultivation zone for blue carbon, and a multiuse zone that has quotas on harvesting. In this third zone, trees can be cut, but at a very limited capacity:

> After a few months, the state decided on laws that totally and strictly prohibit the cutting of mangrove forests, even in the quota in the controlled-use zone. While the project has asked the state to reverse its decision, many of the villagers feel betrayed by the NGO and no longer trust it to protect their interests.[102]

Jean-Robert, a full-time technical worker, noted that the community thinks the project "tricked them" in some way, and it was the state that declared all along that anything under the carbon project was going to be restricted.[103] He said, "We can't do anything against the state. . . . But hopefully we can regain trust between the community and the project." However, Belo, a female employee, noted that, as an NGO, they always work closely with the community. However, the state does not take this relationship into account. It is this that differentiates NGOs and the state: NGOs consider public opinion and they are adaptive in their approach.[104]

This is how the work of science fits back into a local legal justice system officiated by the *dina*. Local-level scientists are now not only monitoring and reporting infractions, they are evidence seekers, and sometimes called as witnesses to testify against other community members. There is also tension, as one person noted, "provoking enemies" among different members when some have to play the role of the "bad cop" on behalf of the project. For this reason, the respondent decided to resign:

> If there is anyone who is doing proven destructive activities, it is we who are doing the enforcement. When the session is over, there is no support coming from the project or the association. We are given a payoff per diem and that's it. No thanks![105]

For many of the villages, this has provoked mixed feelings, sometimes with really perverse outcomes, as the villagers are not only excluded from previously agreed upon extractive zones but are also the ones monitoring and calculating their own demise through ecological surveys. As a former male worker for the project, Anja, explained, "Personally, I am interested and aware of the benefit of blue carbon and especially conservation. For me, working for the association is not a profession; it's a pleasure." However, Anja is also very aware that serious tensions exist:

> The project has made the life of the villagers worse. If we cannot convince the state to reverse its decision, it is difficult because the community thinks that the project and the state lied to them. It is as though we are blocking our own lives, while we manage the project for them.[106]

Rindra, a young-adult male who was formerly a member of one of the associations, explained that, for the state, cutting down *any* mangroves is prohibited. But, he noted, this carbon fund is intended for several social activities such as construction of schools and hospitals, security for beneficiary villages, and so on: "It is for people who are not against the NGO or the blue carbon project—that is to say, for those who respect the *dina*." In these circumstances, one cannot really dissent from the project, or one may be left out when the "goodies" are handed out. Another former member, Julien, speaks in even harsher terms:

> In the association, there is favoritism for family and loved ones, who have priority for benefits. The project has made life worse for people like me, without these connections, now, because before I was able to cut mangrove wood to earn money, but now it is forbidden for me, and my income has been cut off.[107]

On the other hand, others expressed the desperate need for the project to continue and a fear of the state coming back and controlling their livelihoods:

> We thank the project for having created the association. Before their arrival, we had a lot of fears about fishing inspections and extraction in the mangroves. When the Velondriake Association was born and said it was going to protect marine resources, the state no longer came to inspect us. Our children got to school and the project takes care of them all the way to university, if they continue. We can no longer do destructive activities in our area, and if the migrants come here, we forbid him to do that.[108]

The desperation of some of the communities was rather stark, with citizens worrying about their children because, as poor people, they would no longer be able to afford to send their children to school to get an education. They lamented that their work would one day be over, because their sponsors would no longer be able to afford to finance it. Others were really feeling the effects of what had happened with the fishing reserve. A female fisher, Tiana, who was not asked to participate in the work of science monitoring, expressed her dissatisfaction with her situation:

> It is better to live without the project than now, because everything is prohibited with the mangroves. . . . For us, it is with the mangroves that we build our house; even if we always take the risk, we live in anguish of being caught and then paying the fine; you can no longer pass even around this reserve, and the new restrictions greatly disrupt our lives and cause hardship.[109]

Jean, one of the younger fisherfolk, also noted his dissatisfaction with the project because it made him "suffer" by prohibiting mangrove cutting. Another noted: "Mangroves are our source of income, our life, and our livelihood for myself and the entire community," observing that while conservation may be "good in the sea," when taken on land, it falls short.

This ambivalence toward conservation was also felt by Ny, female fisherfolk who liked the idea of the project because there are lots of marine resources that live in the mangrove: "I'm a little worried about the cutting ban, because I don't know what to do, and I don't want to cut my relationship with the project because I'll lose benefits. But if can't build a house . . . I might have to steal mangroves to do it."[110]

Clearly the blanket moratorium on cutting mangroves and the state revenue grab is having effects even on those closest to the project. This outcome,
however, could have been foreseen. It is the outcome of the state and its
relationship to international NGOs, who can mobilize local labor to create
a conservation commodity under difficult ecological conditions and crises.

Conclusion

In this chapter I have discussed how blue carbon, and the blue economy in
general, is yet another attempt to justify conservation access to a host of new
and globally underexploited resources. As capital expands due to crises, it
is forced to seek out new frontier environments to commodify and labor
pools to help it do so. This chapter has demonstrated that even the most
marginal communities are now being ensnared in the deepening and widening of capitalism under new blue carbon initiatives. One needs to consider
what the commodification of these frontier economies and environments
means to those who are now employed to map and measure carbon locally.
High-quality carbon offsets are now being used both to expand market conservation into new frontier areas and to ensnare locals, who can participate
through their labor to add value. Unlike the other two cases in the book,
where local precarious labor is generally hidden and underreported, here it
is openly exposed and promoted.

This does not begin and end with mangroves, however; it encompasses
deep-sea extremophiles and other uncharted frontiers that might hold the
key to novel drug discovery and other biodiscoveries. The acceleration of
"blue growth" has had its own share of critique.[111] Enclosure of massive
swaths of mangrove forests for globalized carbon markets or coastal fishing
and marine conservation and the expansion of the blue economy into coastal
marine and ocean areas are now being questioned.[112] And when adopted,
climate mitigation at the local level usually gets layered onto already existing conservation and development projects, without regard for the already
existing structural inequalities. In spite of the good intentions, in the end,
this may do more harm than good, causing significant friction at the
local level.[113]

The effects of these "grabs" of marine conservation and coastal areas were
felt immediately, especially at the local level and by the most marginal, lead-

ing us to question: What labor regimes emerge from these new climate adaptation and mitigation programs? What are we left with as we move into new carbon markets, with flexible, sometimes invisible and sometimes highly visible labor?

This shift toward the oceans also coincides with increased knowledge of the oceans and advances in technology. We now know more about the oceans than we ever have and also have the technology (such as LiDAR—light detection and ranging, the imaging technology using aerial water-penetrating green light to create seafloor maps—and hydrological surveys) to increase our knowledge. As we learn more about the oceans, we will begin to explore new places to offset carbon and seek out biodiverse areas that may lead to the discovery of new natural products—themes I turn to in the concluding chapter.

Conclusion

Reconsidering Conservation Commodities and the Visibility of Ecologically Precarious Labor

> Such forests are, as it were, the industrial heartlands of nature, where a rich supply of energy mobilizes the earth's minerals and chemicals to make more kinds of products.
>
> *—The Economist*, 1988

A focused alternative to earlier conservation approaches, the "hotspot" conservation strategy targets unique areas of the globe that contain exceptional biological richness, together with threatened species that are in urgent need of protection. As the number of hotspots have expanded across the globe, the strategy has also coincided with a shift by environmentalists away from draconian "fortress conservation" policies, toward new market-based models, promising economic and social development for local communities across the Global South.[1]

Ceremoniously launched at the 1992 Earth Summit in Rio, market conservation was guided through macroeconomic approaches by large organizations, and multilateral and bilateral donor institutions, with an interest in socioeconomic development and debt reduction.[2] The logic behind market conservation was to incentivize national governments and "frontline" local and indigenous people to protect their biodiversity through direct financing and links to global markets. This began with the rollout of new Integrated Conservation and Development Programs (ICDPs) such as bioprospecting,[3] debt for nature swaps, and community-based resource management. It soon advanced to a hybrid mix of traditional extraction and more environmental "service-oriented" programs, such as biological offsetting, REDD+, and payment for ecosystem services.[4] These latter programs were designed essentially to pay governments and local people to leave nature alone.[5]

Over forty years since the Rio event, many of these more service-based conservation programs have been repackaged under the green, blue, and bio-economies. In Madagascar, and elsewhere across the Global South, these frontier "green" economies are bundled as a financial engine for conservation and development, under a veil of the cascading environmental crisis. This matching of interests between the global conservation community and multinationals is, for some critics, transforming "nature under capitalism," and uses the "protection" of parks and protected areas as an accumulation strategy.[6] The spatial widening across the Global South and the industrialization of nature through the interlocking of interests between capitalism and conservation was once *a* way to address the global environmental crisis. It has seemingly become for many even after years of mixed results, *the* dominant way forward.[7]

This book demonstrates that it is under this market rubric that we are witnessing the congruence of bioprospecting, ecosystem services, and mangrove blue carbon projects within the marketing and privatization of nature, and through this process, putting those most vulnerable inhabitants on earth, who rely on these resources, at risk.[8] While this book focuses on case studies in Madagascar, parallels are found throughout the Global South, where the same sites that house some of the most critical biodiversity, targeted for extraction, have been secured under an aggressive use-it-or-lose-it strategy.[9] The production of this conservation commodity frontier, and the tensions it now creates within smallholder communities, raises significant moral, economic, and justice questions surrounding the purported benefits and unforeseen burdens of participation.[10] In particular, this uneven development envelops the mobilization of local-level labor as a way to embed locals into projects, often with perverse repercussions. Running through much of the preceding analysis is the idea that somewhere, hidden in nature, remains the key to solving our environmental problems.

The commodification of nature has been shown to be a tricky business; be it the material characteristics of the resource itself or social and political resistance to it,[11] nature's recalcitrance to be transformed into conservation commodities has led to either imperfect or incomplete markets or environmental services. At one level, this book describes attempts by multiple market conservation groups to overcome such natural, social, and political barriers. It documents, among other things, the professionalization of a local-labor force to count, measure, and map, and to make nature decipherable for market conservation and development, and the opportunities and tensions that arise from this at the local level. I draw off political ecology's attention to complexities of access and control to valuable resources, and in doing so expose much of the hidden labor devoted to making nature eligible for capital accumulation.[12] Insights and critiques from feminist political ecology have significantly contributed to, vastly enhanced, and greatly improved our ability to analyze and set new directions to see and understand resource control and access as a "theatre of contested entitlements."[13] Through this, we can "foster a consciousness of oppositional world views, and [learn to see] differently."[14] Adopting a political ecology lens exposes the hidden labor often promoted as "local participation," and the book foregrounds the role of such hidden skilled and unskilled scientific labor in generating conservation commodities. It is through this lens of labor that I examine the privileged and gendered access to resources and capital of powerful policy actors and government elites, and the implications in terms of rights, equity, and sustainability for those upstream actors in the green commodity chain.[15]

In the final chapter of this book, I illustrate the principal implications for understanding the practice of frontier economies in a period of global planetary crisis. This identifies the paradox at the center of market-based sustainable development: many attempts to commodify nature have not only fallen short of their expectations to save it but upended the sustainable development goals they set out to accomplish. The "silver-bullet strategy" of using market tools represents the ambivalence central to the sustainable development overall, together with its central paradox.

This chapter ends with some suggestions for possibilities, challenges, and alternatives for the Global South.[16] This discussion is meant to open up space for students, scholars, and the public to continue to ask critical questions around the marketization of nature, and the often hidden and precarious labor used to make it legible for the market.

The Seemingly Limitless "Nature" of Commodity Frontiers

Every day it seems we are bombarded with fantastic media stories describing the marketing of nature as a way out of our current environmental predicament. One can see the attraction of these headline stories: turning atmospheric CO_2 emissions into perfume, using mushrooms as a cement, super drugs from sloth hair, and employing spiders' webs to create organic bullet-proof vests.[17] According to this narrative we do not need to change the way we consume, nor do capital and the supporting state structures need to change how they behave; in fact, it gets better (and even more dangerous!) as we are told to believe that our hyperconsumption patterns can actually transform lives and save the environment.[18]

Not only is this turning nature into new products and services, but it is also pushing back the frontiers of where such tradable nature can be found.[19] These ideas range from practical science to science fiction, with plans to find new drugs deep in the hypothermal ocean vents and mine green extractives on the moon.[20]

For instance, not long into the COVID-19 pandemic, there seemed to be an endless number of articles about the potential of new drug compounds that could bring us out of the global health crisis.[21] And who can blame us? COVID-19 was a traumatic and life-altering event, and with its origins all pointing toward zoomorphic drivers through human consumption of wildlife, it just seemed to layer on top of a host of already cascading and linked environmental crises: climate change, species extinction, and deforestation are driving humans deeper into rainforests to seek out food and natural resources.[22]

One story seemed to encapsulate the real sense of severity of the time. The novelist Helen Scales, who wrote an article in the *Guardian* entitled, "Covid Tests and Superbugs: Why the Deep Sea is Key to Fighting Pandemics," laid out in detail the fascinating possibilities of such discoveries and the potential pitfalls to exploring the oceans.[23] The article concerned the bioprospecting of marine life from species found in the most extreme environments, known as extremophiles, organisms that have built up a unique chemical arsenal to withstand difficult ecological conditions such as high salinity, heat, or acid conditions, and thereby may hold novel bioactivity that can be used against disease targets.[24] In the piece, Scales states that "the hit rate for finding powerful and useful new compounds is proving to be especially high among animals of the deep sea." She noted that

> hundreds of biologically active compounds have been found at the bottom of the ocean, some already in widespread use. Enzymes found in bacteria living around hydrothermal vents are even being used in tests for the Covid virus. . . . Even after forty years of scientific research since hydrothermal

vents were first found, a tremendous amount is still being discovered about these extreme ecosystems, which thrive in scorching, toxic waters pouring through cracks in the deep seabed, miles underwater.

The bioprospecting of unique extremophiles is not necessarily a new process. In fact, the story of new COVID-19 drug discoveries is another example of cascading environmental problems. One was also a cautionary tale of the fragility of the deep ocean and our incessant desire to search for hope in elusive natural resources that will somehow take us, and themselves, out of crisis. These environments are also under extreme duress due to new proposals for deep-sea mining for critical minerals used in the green extractives for renewable energy and battery storage, as well as the search for new strains of antibiotics. This is because "mining companies want access to the seabed beneath international waters, which contain more valuable minerals than all the continents combined."[25] Scales describes the threat: "Yet novel antibiotics and an untold variety of beneficial molecules could easily be wiped out if ecosystems around vents and elsewhere on the ocean floor were to be destroyed by deep-sea mining, which could go ahead in less than two years."

If these use-it-or-lose-it justifications for getting out there and capturing what we can before it has all gone seem familiar, that is because they are. Similar arguments were also made to justify the collection of rare and unique biodiversity from Madagascar's rainforests thirty years ago. It was not only the uniqueness of these species but also their risk of extinction that necessitated both their collection and identification, followed by commercial production. However, rather than resisting large-scale extraction for the sake of drug discovery and environmental protection, conservation organizations embrace it. It is as if nature had to begin to do its part, begun to pay for its own survival.[26]

This anecdote is relevant to the closing discussion of this book in two distinct ways: first, it speaks to the widening and extending of the commodity-centered market conservation under crisis.[27] Rather than a critical reflection on our relationship to nature, we double down, both intensifying and extending nature's industrialization, financialization, and capture.[28] We will, it seems, get to a point where there is nowhere left to search, nowhere left to dig, and no way left to invest our way out. Second, as these new commodity frontiers are discovered, or reintroduced in the media and policy discourse, we begin to notice the relayering and rebundling of old tropes into new ones, causing frictions and social and political tensions wherever they touch down.[29]

These bundled nexuses of commodity frontiers have become a focus of study by geographers and critical political ecologists. Political geographer Philip Le Billon describes the extractive industries' attempts as showing the deepening and widening of how "conservation has become an intrinsic part of the ways many extractive companies portray themselves in the context of the contemporary 'extinction crisis.'"[30] This is beyond the act of corporate greenwashing; it is *the* way that conservation is now guided.[31] These intersections cannot be overlooked, from ecotourism[32] to bioprospecting[33] and carbon credit programs.[34] In the near future, we can only expect more of the same relayering and rebundling through the articulation of land/sea dynamics[35]—for instance, policy discourse developing around deep-sea offsetting.[36]

It is no accident that this layering of politics, science, and ecological breakdown seems to cluster together so neatly. The nexus of commercial drug discovery and intersections of deep-sea mining in fragile marine ecological zones is, in effect, a perfect representation of a commodity frontier—a development imaginary of nature's commercialization under crisis. It is a deepening and widening of both where and how capital acts on nature. Capital is not alone in this story, however. Someone must do this work, carry out the actual labor, not only to create policies and permits but also to carry out collection and discovery. The need for this work and the people who carry it out are often obfuscated. This book is about talking through this conservation and development work as it exposes the paradox of the green, blue, and bio-economies.

The Hard and Sometimes Hidden Work That Goes into Making Nature Legible for Market Conservation

If anything, this book has demonstrated that conservation and development is hard work. From the boardrooms of multinational firms to policy meetings of development donors, it is clear that significant time and effort is put into locating and delineating land and nature for climate programs, and wider green, blue, and bio-economies.[37] Yet less understood is the local labor that is equally vital to conservation and development. While generally seen as

Figure 16 Malagasy weighing biomass for carbon potential in mangrove forests in southwestern Madagascar. (Photo by Garth Cripps)

separate, this triple-nexus of sustainable development is now more and more enmeshed in the sociocultural and ecological fabric of those hired to carry out this work. From porters and tree planters to those counting lemurs and measuring tree density in carbon landscapes, the way labor is envisioned in these settings has similarities with the much-talked-about gig economy. Workers in the gig economy were first identified in the postindustrial settings of the Global North and are considered to be flexible, disposable, underpaid, and invisible.[38]

The three case studies in this book represent some of the most progressive and successful examples of the green, blue, and bio-economies. The first is a historical case study, describing one of the longest-running U.S. federally funded bioprospecting projects in Madagascar—the ICBG. The project was heralded as delivering on many of its promises of small-scale sustainable development in local areas and training hundreds of Malagasy botanists and pharmacologists, while delivering training and equipment to many others.[39] Historically, Madagascar has been, and continues to be, a destination for bioprospecting programs, and its capacity to contribute to pharmacology and botany can be attributed to the ICBG.

The ICBG also set the foundation, in one respect, for how large-scale market-based conservation and development could operate in the future, through local-level associations set up to organize and carry out the day-to-day delivery of small-scale economic and social development in villages. For bioprospecting collection to take place, local labor was needed to set up base camps and provide guides, cooks, and guardians. Such collection does not take place without the delivery of compensation packages. This was organized by newly formed conservation associations—a sort of precursor to the *dina* agreements in the latter two case studies. Bioprospecting therefore provides a historical framework, indicating how conservation and development would unfold, through local-level organizations that would act as proxies for the organization and delivery of benefits.

Even so, research conducted in the early stages, when knowledge of bioprospecting activities was freshest in the minds of those who participated, was close to nonexistent. Not only did many of those who helped scientists carry bags and collect materials know little about their activities for making new drugs from nature, but they had very little knowledge of the fact that they were supposed to be compensated for the larger objectives of the project. Later research did show benefit-sharing was delivered to some rural communities through upfront compensation; yet given the time-lapse between drug discovery and collection, it is questionable how much of the link between bioprospecting and compensation was made, and if any type of local-level conservation can be attributed to the drug discovery process for those locals who did help out in the collection process.

In the second study, the Ambatovy nickel and cobalt mine is observed as a model for sustainable mining. Its groundbreaking offsetting program was seen by many in the industry as a way forward for extractive industries operating in areas of high biodiversity. The mine is the largest Foreign Direct Investment (FDI) operation in the history of Madagascar, supporting thousands of jobs in a country with few sources of revenue generation. Offsetting seems to have been shown to achieve its environmental objectives and is on track to deliver the no net loss of forests set out by the mitigation hierarchy.[40] Yet beyond the mine's official pronouncements, the social objectives have been less successful.[41] The mine has restricted access to livelihood resources and has provided income-generating activities to compensate the locals, but these have been shown to be generally insufficient and ineffective for those who rely on shifting agriculture, and who can no longer access the forest for a multitude of things including food, fuel, and construction materials. Research has demonstrated that the negative effects of the mine have far outweighed the livelihood substitutes, posed as benefits of these new conservation restrictions.[42] In short, compensation just hasn't cut it, mainly due to the time lag between the immediate lack of access and when benefits were set to be delivered. While this seems like an easy fix, unfortunately, for those who depend on such immediate resources, it's a matter of life and death, with those more vulnerable absorbing the brunt of the offset establishment.[43]

However, the mine has done more than just reassemble the livelihoods of those living closest to the offset areas; it has also changed how they interact with each other. Building on earlier success in developing local-level associations, the mine mobilized a rules-based system called *dina*, which helped to transform the locals' relationships to forest resources. No longer were these livelihood resources but rather things that needed measuring and monitoring and, therefore, training as to how to report back to the mine not only about the forest but about *who* was taking *what* out of it. Prospects look hopeful for the mine operators, as they see the upsurge in demand for nickel and cobalt. However, what will the prospects be for those living closest to the mine operations? They seem to be getting squeezed from both ends, with the mine and conservation offset enforcement pushing vulnerable populations to migrate and attempt to survive further from the mine site.

Lastly, organizations promoting blue carbon have been lauded by the international environmental community for their success in bringing to light the often-hidden importance of mangrove ecosystems and the communities that rely on them. Alongside this is the intricately linked conservation work around fisheries that these emerging carbon markets are meant to support. They have demonstrated a distinct way to finance marine conservation and the potential for community-based associations to maintain control over their resources. However, even with all the rhetoric around expanding the potential of national economies into the ocean frontier, the case of blue carbon demonstrates that capitalism always needs to touch down onto land somewhere and wrestle with the local- and national-level politics.

While the NGO managing the Tahiry Honko blue carbon project has had quite a bit of success in its attempts to deliver marine conservation and mobilize locals around fish stock management, its work in developing and delivering a sustainable blue carbon market has illuminated the shortcomings of relying on global markets for local conservation. As with most conservation projects in the Global South, the potential for these projects to bring in revenue is noted and appropriated by cash-strapped governments searching for sustainable financial streams for protection and management. This "revenue grab" coincides with resecuritization of the natural resources by the state, and a concomitant decrease in access, and thereby livelihood repercussions, at the local level. Indeed, this case of the carbon markets from mangroves demonstrates just that: as mangroves began to bring in revenue, the state stepped in with a moratorium on all mangrove cutting in order to squeeze out whatever revenue was available.

Mangroves have proved yet again that nature's materiality matters. Embedding the project in existing carbon mitigation schemes of REDD+ essentially gave the state jurisdiction over managing the revenue that flowed in from the project. However, in order to make this blue carbon a "high-quality" carbon offset, community participation and labor was promoted and articulated in the selling of the commodity itself. Unlike the hidden labor in previous projects, this work was therefore transformed by the project into "community participation." As we see, the community therefore is tied to blue carbon by their labor, mainly through measuring, mapping, and, most noteworthy, enforcing protection of mangrove sites. However, as enforcement became "localized," and the state tightened its screws on access quotas, this squeeze made maintaining some sort of existence alongside the project, for some, untenable.

Related to this is a second major outcome of blue carbon, similar to the two other cases in the book; that of the creation and mobilization of a pool of local labor, which is in fact a group of precarious workers kept on hand to pick up essential, albeit flexible, labor tasks. These eco-precariat do the hidden work of conservation and development. They conduct an array of unskilled work, guide researchers, carry equipment, guard field sites, and cook food. However, they also do semiskilled work, which is vital to make nature real for the market. This work includes supporting the scientific work in counting fish stocks, crabs, lemurs, mangrove tree stumps, and other endemic species. They are also counting their friends and neighbors (and sometimes themselves), who knowingly or unknowingly might be extracting livelihood resources out of newly protected blue carbon sites; in effect, they are now monitoring their own demise as the tightening and sanctioning of extraction quotas has caused significant hardship adjacent to blue carbon sites.

In some respects, one might say that conservation and development has come a long way in the past fifty years; in other ways, not much has changed.

While local and indigenous populations are now becoming more central in decision-making processes (as a way to bring in local participation), some familiar patters begin to emerge, especially relating to how and in what ways these voices are mobilized. As many have discussed before, participation in sustainable development is never politically neutral. In fact, local participation can cause serious tension, exacerbating long-standing tensions on multiple scales, from gender politics in the domestic sphere to village-level, regional, and ethnic politics, and even contentious border disputes across states.

Yet, like the other case studies of bioprospecting and biodiversity offsetting (highlighted in chapters 3 and 4, respectively), blue carbon in Madagascar relies on both skilled and unskilled labor to make it work. This labor participates in monitoring stocks, surveying mangrove tree-felling for charcoal, and collecting data on carbon and associated fish stocks. We have returned to bioprospecting in this conclusion to discuss how it is used to justify access to and conservation of deep-sea extremophiles and, more commonly, corals and organisms promoted in the "blue economy" discourse, as uncharted frontiers holding the key to novel drug discovery and other bio-discoveries.

At least early on, all three of these cases have attempted to use a decentralized network of decision-making and rules-based resource management, based on local input, or *dina*. Albeit not perfect, they represent a departure from previous command-and-control approaches of state-mandated conservation (fences and fines approaches), which tended to set up areas with few local resource rights. Many now recognize that, with the intensification of market-based approaches, there is the potential for new revenue generation and thereby a securitization and reterritorialization of natural assets. In parallel, we are also seeing a return to "fortress conservation" approaches. Counter to this there are also new approaches discussed below, and my hope is that future academic studies, and the wider public, can pick up on this work and ask these questions moving forward.

Future Studies

One of the main motivations in writing this book was to provide a contemporary account of how local labor is transformed under market conservation. It has been written so that scholars and practitioners can learn both from past mistakes and conservation successes. It has also been written for those working in broader policy-making, and members of the public trying to understand and design more progressive and inclusive strategies beyond the market, around the sticky and sometimes intransient politics of nature conservation. Furthermore, while the case studies of the book focus on Madagascar, it is clearly relevant for many high-biodiversity areas in the Global South (and even in the Global North), where political, ecological, and social tensions around issues of access and control, gender, and labor emerge with the commercialization of nature under crisis.

Conceptually, the eco-precariat can be applied as a theoretical lens beyond conservation and development interventions in the Global South. In fact, it has wide-reaching relevance, especially with emerging scholarship engaging in labor and just transitions, low-carbon economies, e-waste, and green energy.[44] Eco-precariat labor is also highly relevant to studies on climate adaptation. As extreme weather and hazardous environmental conditions become more frequent and widespread due to climate change, there will be an increased dependency on emergent, sometimes hidden, cheap, and flexible labor to deal with both the aftereffects and to prepare for the next round of extreme weather events. Future studies may benefit from thinking through how the eco-precariat, in its many variegated forms of labor, can be applied within different empirical contexts concerning climate adaptation.

The book provides a point of departure for students of political ecology developing critical research questions in feminist political ecology and precarious labor. For this purpose, I propose below some very brief annotations of current threads ripe for future engagement, alongside some of the lessons that can be learned from the book's findings. Examples listed below are not an exhaustive list, nor will one find an exploration of the rich debates and emerging critiques of these approaches; rather, they provide just a few potentially important pathways for scholars and others to take going forward.[45]

Postdevelopment and Conservation Commodities

A significant and rich literature has sprung up in recent years around the idea of postdevelopment alternatives: for instance, research delving into decolonial models seeking to rethink development paradigms through the lens of local and indigenous people, while also recognizing the deep structural and racial aspects embedded in colonial development ideologies.[46] While this is by no means new, critical scholarship has been linking the modern conservation and development issues with neocolonial "domination, oppression, and exploitation."[47] Political ecologists such as Ariadne Collins and colleagues have called out the current conservation and development paradigms as reflecting not a new event but an "ongoing coloniality," extending from, and intensifying, uneven colonial relations. Elsewhere, Tanzanian environmental social scientist Mathew Mebele and colleagues express that, although the term "decolonization" has gained traction recently, it is rooted in historical processes that need to be addressed through conservation education, from that of policy actors to civil society and the broader public,[48] together with focusing on the role of capitalism and enrollment of participatory subjects (e.g., locals).[49] This is clearly relevant to the broader approaches taken in this book of centering capitalism and the market in historical and current approaches of conservation, but unlike a decolonization approach, my focus on labor and conservation commodities problematizes essentialism, observed in new "participatory" conservation approaches.

Critical political ecologist and climate justice scholars Farhana Sultana and Paige West,[50] among many others, have taken up similar calls for building networks of more gendered/intersectional, indigenous, and local engagement across critical development studies, especially studies of climate justice.[51] At the forefront of this work is the criticism of feminist scholars doing political ecology work. Pointing out the inherent violence in asking, "Are We Green Yet?" Wendy Harcourt and Ingrid Nelson argue for a feminist political ecology rooted in Donna Haraway's idea of situated knowledge; acts of imagining "green" or "just" futures come from the privileges, status, and other features of the individual or community doing the imagining. They ask:

> Who within particular environmental, feminist and justice movements asserts which imagined futures? Whose voices are silent or silenced in these visions and goals? The elephant in the room is not who has the "agency" to speak but who has the authority to speak—Global South, Global North, young, old, woman, man, white, black?[52]

Linking market conservation to colonialism and the power of decolonial discourse has significant potential. How does this materialize into resistance movements, new social and direct action, and/or new radical approaches to conservation and development beyond the market? Clearly, paradigms surrounding decolonial and climate justice, and those of feminist political ecology, have gained traction, as Sultana notes; once on "the margins," these issues are now front and center in critical academic discourse.[53]

In response, climate justice scholars have also taken up the charge, not just to see climate change as a global problem but to look critically at those deep racial and intersectional fault lines, beyond the singular and myopic lens of class and toward race, gender, sexuality, and ethnicity. All intersect, and the impact of climate change has the biggest effect. Globally, climate justice scholars have also sought out retribution for those countries who are at the heart of climate emissions in the Global North, and it is important to duly compensate those countries who now must bear the cost of the unfettered fossil fuel–laden development.[54] Students can think hard about how intersectionality can be woven into useable frameworks to critique and also build. This is what political ecologist Paul Robbins calls "hatchet and seed"—a way to build upon critique to provide equitable outcomes over contested resource access and control.[55] How can we begin to develop critical questions that lead to new models of real inclusion and multiple and pluralistic forms of environment and climate justice?[56]

In terms of alternative conservation, scholars have investigated new directions for a more inclusive and socially just practice. One such approach, convivial conservation, attempts to take on the practice of conservation through a lens of equal coexistence and management and "confronts the structural, violent and uneven socio-ecological pressures of our current economic system."[57] This approach, according to its architects, political ecologists Bram Buscher and Robert Fletcher, helps to ensure that convivial conservation will emerge:

> [It is] very different from current practice, namely a use of parts of nature that is sustainable (i.e., not geared towards eternal quantitative growth and accumulation), whilst being part and parcel *of* nature. It would entail *living with* other aspects of nature in ways that balances human and nonhuman needs. Indeed, conservation itself would be integrated and (re)embedded within daily life and all other domains of policy and action rather than something we do mostly in protected areas or when donating to an NGO. Moreover, convivial conservation moves away from capital-inspired ways of "rendering visible" the value of nature, and instead becomes a part of broader structures of democratically sharing the multidimensional wealth that nature embodies.[58]

So how does this conservation differ from current discourse around inclusive or participatory conservation? Can it be scaled up in a way that resource-strapped countries in the Global South may find more attractive than, say, revenue that can be generated from green or blue economy platforms, such as biodiversity offsets? What are the divisions that may begin to form when a less market-based and more convivial conservation begins to emerge in biodiversity hotspots? Finally, what forms of labor begin to emerge, and disappear, when the market is neutralized in such a way?

Another radical perspective includes that of degrowth.[59] Taken from the French *décroissance*, the term "degrowth" designates an approach that is critical of the hyperconsumption in the Global North. The movement has galvanized support in providing an alternative to capitalism and a way for eco-socialist livelihoods to flourish. Jason Hickel defines degrowth as "a planned reduction of energy and resource use designed to bring the economy back into balance with the living world in a way that reduces inequality and improves human well-being."[60] Kallis defends degrowth as a "a potent political vision that can be socially transformative, including . . . a full ensemble of environmental and redistributive policies [being] required, including—among others—policies for a basic income, reduction of working hours, environmental and consumption taxes and controls on advertising."[61]

It is a radical transformation of society revolving around a set of critiques of the capitalist paradigm, emphasizing the reduction of global consumption and production and seeking alternative measurements of development beyond outdated metrics of Gross Domestic Product (GDP). Indeed, it advocates social justice and societies who hold more accountable indicators for ecological sustainability and social and environmental well-being.[62]

These alternative approaches also align with other approaches taken from the Global South, such as the commonly highlighted Southern and Eastern African approach *"Ubuntu"*—a perspective that seeks to build on community enhancement rather than individual,[63] or the Bolivian and Ecuadorian perspective of *"Buen Vivir"* (good life). *Buen Vivir*, as enshrined in Ecuador's 2008 political constitution and National Development Plans, seeks a plurality of voices advocating alternatives to the extractive and capital-intensive development that has shaped many countries across the Global South.[64] But how does one transition to a scaled-up degrowth economy in the Global South? Delinking from the global economy seems intractable, while the drivers of hyperconsumption from the Global North continue to increase. Can Global South economics provide a model for how we are to rethink our existence in the North?

For some, small may be beautiful, but it may just not be enough. Some prominent voices currently hold views that, given the predicament in which humanity currently stands, broadening perspectives on new technologies and approaches, such as geoengineering biotechnology, gene-editing, or enhanced algorithmic machine-learning (AI), is the right solution to divert us away from a climate breakdown.

> And what is to say that local conservation methods are any better at actually conserving biodiversity? Clearly, any conservationist who has worked in Madagascar sitting on the side of an upland tavy rice field would argue otherwise. And it would be difficult not to agree. It is about the relations between those enforcing conservation and populations who will bear the brunt.[65]

While readers of this book might have to look elsewhere for full critiques and well-trodden debates around each of these alternative perspectives, I set forth some new questions for students and scholars alike, who are looking to see where these perspectives are, in practice, touching down, to enable them to intersect with green, blue, and bio-economies. Where does labor get mobilized and hidden in carrying out these new perspectives and proposals, and who are the winners and losers in this new labor force?

Responses from Malagasy Scientists

Moving forward, one might ask, what does the future hold for conservation across the Global South? There is no doubt that if we look at narratives coalescing around global environmental conditions, the future certainly looks bleak. Continuing deforestation, food insecurity, and climate change challenge both the status quo of markets as the savior as well as a long-held and somewhat counter belief that if locals were able in some way to participate in the creation and development of conservation, then a more just and equitable outcome might emerge. Conservation outcomes that are not necessarily solely dependent on global markets or commodity-led development may actually improve the livelihood of smallholders, and they would no longer face the brunt of these multiple global ecological crises, nor the shortcomings of policy solutions available to support them.

So, what if we have it all wrong? What if intensifying the market is not the way forward? Where do we go from here? Some Malagasy researchers I have spoken to have a more optimistic outlook. In fact, speaking to a number of Malagasy scientists about what the future holds actually suggests a more hopeful future, and many have already observed a transformation in how conservation and development policy is created and rolled out, leading to a different picture for conservation in the next five to ten years.

There has been a rise of the Malagasy female scientists through the conservation community in Madagascar, and their approaches to conservation and the way they, and many of the civil society and private sector institutions and organizations, interact with local communities and the state. This was noted by one experienced Malagasy biologist who leads a large internationally funded project in Madagascar:

> Younger scientists trained in a Western country are more activist, and NGOs have begun to reflect this change, or the desire to change. For in an era of climate change and new technology and media, the way in which the NGOs act must change.
>
> For years, Madagascar has been conducting strict conservation, like getting people out. Yet we have the new generation of scientists; we try really to think about locals' livelihoods. We have to be clear with locals and NGOs that if they want this kind of a development, that we need to develop an understating. We now need the state to take responsibility in enforcing conservation, . . . to use standard international standards and use law enforcement correctly, not too harshly.[66]

Another female scientist, who also works in a private sector program in offsetting and conservation, noted:

> From what we are currently doing, especially in terms ecological restoration and offsetting, I think we have been very successful in Madagascar. We are quite advanced and everyone wants to learn from what we are doing—both in Madagascar and abroad. They want to hear us speaking more, and bring more of our methods to the forefront. Yeah, I am very proud of this. I have to say that we Malagasy are more motivated to do conservation activities,

and I attribute that to our work, which is incredibly prideful. So yes, there is an individual pride in my work when I speak about it.[67]

Her close female colleague, also very high up in a leadership position, noted:

You know, in not only nature conservation and NGOs but also in the private sector, up to this point we did not have many high-level Malagasy women scientists in positions of power. Actually, two of the three big NGOs now have Malagasy women scientists in top leadership positions. Moreover, the Malagasy minister of environment is a women scientist. Those are examples of how Malagasy people are skilled people who can handle large management programs that need the scientific skills together with management skills. The fact that you can make decisions according to science is very important. That is very helpful in terms of that the future of young Malagasy scientists.[68]

One common critique that social scientists often receive is that their data analysis results in conclusions that are naive and do not tell the whole story. It is true that, as with any studies, some tend to essentialize perspectives "from below," they are far from naive. The role of the social scientist here is to provide critical perspectives from less-heard voices. One might call us glorified investigative journalists, but in fact we are trained in techniques that identify hyperbole and exaggeration, and we conduct critical discourse at multiple levels. We see politics of nature in places where many miss it, and imbalances in social relations across multiple scales and spaces. We carry out important methodological techniques, such as triangulation, interviews, focus groups, and mixed methodologies including qualitative, quantitative, and geospatial techniques. However, while some Malagasy and other nationals across the Global South have been trained in social science techniques, many are still only receiving natural science training. This, however, is beginning to change.

Respondent: We have Malagasy who are now twenty years on, twenty-five years on, trained and ready to take over.

Me: Are we ten years away from when Malagasy NGOs are completely run and operated by Malagasy?

Respondent: You know, it is an interesting question. But I was thinking like three to five years, five years.

Me: Cool!

Respondent: There are so many younger women right now that are very, you know, kind of very motivated to be a conservationist, to be a critical leader. I think so. Therefore, I was thinking like five years they would be taking over.[69]

The development of "homegrown" leaders in conservation has been supported by years of training and development for research in the country. One leading U.S. biologist noted:

The future of a country such as Madagascar depends upon nationals, both scientists but also local conservation leaders. Early on there were Malagasy scientists to do conservation, but most were in the shadows of foreign counterparts, more or less carrying out field assistance rather than actively running their own projects. And it was clear for Madagascar as a country to advance in everything associated with development, conservation, resource utilization, etc., it was obligatory to help with the creation of generations of people, nationals that would fill the roles to advance those programs. Therefore, in simple terms, that is what twenty-five years of training was all about, for them to finally take over.[70]

Well, it seems that this idea has finally come to fruition. It's time to hand over the reins of international conservation and development funds to young Malagasy scientists in a way that represents the diversity of gender and the broader diversity of the Malagasy people, beyond just those who can access higher education in the capital and a few select coastal cities.

My only response is to keep on doing the hard work, the critical work, of pointing out when projects fail, and stating explicitly why. As my PhD adviser said, "Keep up the good fight!" That fight, however, takes many forms on multiple levels and through multiple voices. There is no silver bullet. However, there are definitely better options than the ones we have now, and if this book tells us anything, it is that many of those lie somewhere outside the market and in the hope that the younger generation of Malagasy scientists has suggested.

Notes

Introduction

1. R. A. Mittermeier, N. Myers, J. B. Thomsen, G. A. Da Fonseca, and S. Olivieri, "Biodiversity Hotspots and Major Tropical Wilderness Areas: Approaches to Setting Conservation Priorities," *Conservation Biology* 12, no. 3 (1998): 516–20; N. Myers, R. A. Mittermeier, C. G. Mittermeier, G. A. Da Fonseca, and J. Kent, "Biodiversity Hotspots for Conservation Priorities," *Nature* 403, no. 6772 (2000): 853–58.

2. N. Myers, "The Biodiversity Challenge: Expanded Hot-Spots Analysis," *Environmentalist* 10, no. 4 (1990): 243–56.

3. Myers et al. "Biodiversity Hotspots"; R. A. Mittermeier, G. P. Robles, M. Hoffman, J. Pilgrim, T. Brooks, and C. G. Mittermeier, *Hotspots Revisited* (Mexico City: Cemex, 2004).

4. Madagascar currently sits in the Indian Ocean roughly 400 km off the east coast of southern Africa, separated from Africa by the Mozambique Channel. At its widest point, it measures nearly 580 km and is crossed by the Tropic of Capricorn near the southern city of Toliara (Tuléar). It has an area of 587,045 km² (roughly the size of France), and it runs approximately 1,600 km north to south. Its physical topography is sometimes described as "wedge-like," with a string of former volcanic mountains running the length of the island. The highland plateau, which begins in this northern massif, runs along its "backbone," forming a stark climate and hydrologic division between the eastern and western regions. While the east is characterized by heavy rainfall, which can exceed 3,000 mm a year, the west coast is "desert-like," receiving less than 400 mm annual rainfall. The island is also hit by frequent cyclones that cause major damage to farms and road infrastructure.

5. L. Wilmé, S. M. Goodman, and J. U. Ganzhorn, "Biogeographic Evolution of Madagascar's Microendemic Biota," *Science* 312, no. 5776 (2006): 1063–65.

6. S. M. Goodman and W. L. Jungers, *Extinct Madagascar: Picturing the Island's Past* (Chicago: University of Chicago Press, 2021).

7. Madagascar is also said to have extremely high endemism rates at the top-end genus and family taxonomic levels. J. Ganzhorn, U. Porter, P. Lowry, G. E. Schatz, and S. Sommer, "The Biodiversity of Madagascar: One of The World's Hottest Hotspots on Its Way Out," *Oryx* 35, no. 4 (2001): 346–48; J. U. Ganzhorn, L. Wilmé, and J. L. Mercier, "Explaining Madagascar's Biodiversity," in *Conservation and Environmental Management in Madagascar*, ed. I. R. Scales (London: Routledge, 2014), 41–67.

8. Tropical islands, such as those in the Indian Ocean region, have for years been a target of some of the world's first state-led conservation interventions, starting in the late 1700s with soil conservation. R. Grove and R. H. Grove, *Green Imperialism: Colonial Expansion, Tropical Island Edens and the Origins of Environmentalism, 1600–1860* (Cambridge: Cambridge University Press, 1996).

9. Formed by long-term processes, chemical and physical weathering, and prolonged periods of wet and dry seasons common in the tropics, Malagasy soils also lack the same parent material of temperate soils. As glacial deposits receded in more temperate regions of the United States and Europe, they left nutrient-rich rock deposits. Yet most of sub-Saharan Africa (excluding parts of South Africa) missed out on these recent ice age events. P. Roederer and F. Bourgeat, "Carte des sols de Madagascar au 14000000è," in *Atlas de Madagascar* (Antananarivo: Université de Madagascar, 1971).

10. Not to say that environmental degradation is a myth in Madagascar. Historical accounts of deforestation, species extension, and poor air quality in urban areas are only a few of the environmental challenges facing the island. See G. Feeley-Harnik, *Green Estate Restoring Independence in Madagascar* (Washington, D.C.: Smithsonian, 1991).

11. This intensified after shots, in a National Geographic exposé, of the mouth of the river Baeteboka after seasonal runoff, which showed the red soil silting up the mouth of the river. See G. M. Sodikoff, *Forest and Labor in Madagascar: From Colonial Concession to Global Biosphere* (Bloomington: Indiana University Press, 2012).

12. E. Keller, *Beyond the Lens of Conservation: Malagasy and Swiss Imaginations of One Another*, vol. 20 (New York: Berghahn Books, 2015); J. C. Kaufmann, "Introduction: Recoloring the Red Island," *Ethnohistory* 48, no. 1 (2001): 3–11.

13. C. A. Corson, *Corridors of Power: The Politics of Environmental Aid to Madagascar* (New Haven: Yale University Press, 2016); K. S. Freudenberger and the International Resources Group, *Paradise Lost? Lesson from 25 Years of USAID Environment Programs in Madagascar* (Washington, D.C.: U.S. Agency for International Development, 2010).

14. L. Odling-Smee, "Conservation: Dollars and Sense," *Nature* 437, no. 7059 (2005): 614–17, 614.

15. S. Laird and R. Wynberg, *Access and Benefit-Sharing in Practice: Trends in Part-nerships Across Sectors* (Montreal: CBD Technical Series/UNEP, 2008).

16. See B. Neimark, "Green Grabbing at the 'Pharm' Gate: Rosy Periwinkle Pro-duction in Southern Madagascar," *Journal of Peasant Studies* 39, no. 2 (2012): 423–45.

17. For more on the history of the practice of ethnobotany, see https://achievement .org/achiever/richard-evans-schultes-ph-d/. The famed Harvard ethnobotanist Richard Evans Schultes was once known as the "father of modern ethnobotany." He was director of Harvard's Botanical Museum and was recognized as estab-lishing ethnobotany as a universally recognized academic discipline.

18. All dollar amounts given refer to U.S. dollars.

19. B. Neimark, S. Osterhoudt, L. Blum, and T. Healy, "Mob Justice and 'The Civi-lized Commodity,'" *Journal of Peasant Studies* 48, no. 4 (2021): 734–53, https:// doi.org/10.1080/03066150.2019.1680543.

20. K. McAfee, "Selling Nature to Save It? Biodiversity and Green Developmental-ism," *Environment and Planning D: Society and Space* 17, no. 2 (1999): 133–54; S. Sullivan, "Nature on the Move III: (Re)Countenancing an Animate Nature," *New Proposals: Journal of Marxism and Interdisciplinary Inquiry* 6, nos. 1–2 (2013): 50–71. See also B. Büscher and R. Fletcher, "Accumulation by Conser-vation," *New Political Economy* 20, no. 2 (2015): 273–98.

21. McAfee, "Selling Nature to Save It?" See also C. Corson, K. I. MacDonald, and B. Neimark, "Grabbing 'Green': Markets, Environmental Governance and the Materialization of Natural Capital," *Human Geography* 6, no. 1 (2013): 1–15.

22. H. Lovell, "Climate Change, Markets and Standards: The Case of Financial Accounting," *Economy and Society* 43, no. 2 (2014): 260–84, https://doi.org/10 .1080/03085147.2013.812830.

23. P. Bigger and M. Robertson, "Value Is Simple. Valuation Is Complex," *Capitalism Nature Socialism* 28, no. 1 (2017): 68–77. See also J. W. Moore, "The Capita-locene Part II: Accumulation by Appropriation and the Centrality of Unpaid Work/Energy," *Journal of Peasant Studies* 45, no. 2 (2018): 237–79.

24. See, for example, B. D. Neimark and B. Wilson, "Re-mining the Collections: From Bioprospecting to Biodiversity Offsetting in Madagascar," *Geoforum* 66 (2015): 1–10.

25. I am aware that this is a somewhat crude approach to the complex division of workers, but I find it helpful as a starting point. Furthermore, I want to note that these groupings are by no means static, as some move across the group boundaries, drop out, and actively join as opportunities appear.

26. G. Standing, *The Precariat: The New Dangerous Class* (London: Bloomsbury Academic, 2011).

27. More on this in chapter 1. For a full description of the eco-proficians and eco-precariat, see B. Neimark, S. Mahanty, W. Dressler, and C. Hicks, "Not Just Participation: The Rise of the Eco-precariat in the Green Economy," *Antipode* 52, no. 2 (2020): 496–521.

28. Neimark et al., "Not Just Participation," 4.

29. J. W. Moore, *Capitalism in the Web of Life: Ecology and the Accumulation of Capital* (London: Verso Books, 2015); N. Klein, *The Shock Doctrine: The Rise of Disaster Capitalism* (New York: Macmillan, 2007).

30. B. Neimark, S. Mahanty, and W. Dressler, "Mapping Value in a 'Green' Commodity Frontier: Revisiting Commodity Chain Analysis," *Development and Change* 47, no. 2 (2016): 240–65.

31. D. E. Rocheleau, L. Thomas-Slayter, and E. Wangari, *Feminist Political Ecology: Global Perspectives and Local Experience* (London: Routledge, 1997); R. Schroeder, *Shady Practices: Agroforestry and Gender Politics in the Gambia* (Berkeley: University of California Press, 1999); R. Elmhirst, "Introducing New Feminist Political Ecologies," *Geoforum* 42, no. 2 (2011): 129–32; A. J. Nightingale, "A Feminist in the Forest: Situated Knowledges and Mixing Methods in Natural Resource Management," *ACME: An International Journal for Critical Geographies* 2, no. 1 (2003): 77–90; J. Sundberg, "Feminist Political Ecology," in *The International Encyclopedia of Geography: People, the Earth, Environment and Technology*, ed. D. Richardson (Wiley-Blackwell, 2016), 1–12.

32. First quoted to me by my master's committee adviser, Dr. Normal Uphoff, professor of government at Cornell University.

33. B. D. Neimark, "Biofuel Imaginaries: The Emerging Politics Surrounding 'Inclusive' Private Sector Development in Madagascar," *Journal of Rural Studies* 45 (2016): 146–56.

34. J. Moore, "Industrial Revolution II," jasonwmoore.wordpress.com, July 4, 2013, https://jasonwmoore.wordpress.com/tag/industrial-revolution/.

35. J. Fairhead, M. Leach, and I. Scoones, "Green Grabbing: A New Appropriation of Nature?" *Journal of Peasant Studies* 39, no. 2 (2012): 237–61.

Chapter 1

1. *Guardian*, "Rabbi, 77, Arrested at Extinction Rebellion's Bank of England Protest—Video Report," October, 14, 2019, https://www.theguardian.com/envi ronment/video/2019/oct/14/rabbi-77-arrested-at-extinction-rebellions-bank-of -england-protest-video-report.

2. Rabbi Jeffrey Newman, Emeritus Rabbi of Finchley Reform Synagogue, UK. It was the start of the Jewish festival of Sukkot, and the Rabbi was holding a *lulav* (date palm frond) and *etrog* (citrus), two symbols of the holiday meant to represents thanksgiving for a natural harvest season, fertility of land, and desire for rain.

3. D. Gayle and M. Taylor, "Extinction Rebellion Activists Arrested at Bank of England Protest," *Guardian*, October 14, 2019, https://www.theguardian.com /environment/2019/oct/14/extinction-rebellion-activists-stage-protest-at-bank -of-england.

4. G. Thunberg, A. Taylor, et al., "Think We Should Be at School? Today's Climate Strike Is the Biggest Lesson of All," *Guardian*, March 15, 2019, https://www.the guardian.com/commentisfree/2019/mar/15/school-climate-strike-greta-thunberg.

5. Forbes estimated BlackRock's net worth just two years later, in 2021, at $9.6 trillion. See Trefis Team, "With $9.5 Trillion In Assets, Is BlackRock Stock Fairly Priced At $910?" *Forbes*, August 19, 2021, https://www.forbes.com/sites/great speculations/2021/08/19/with-95-trillion-in-assets-is-blackrock-stock-fairly -priced-at-910/?sh=431fff895b5b.

6. *New York Times*, "BlackRock C.E.O. Larry Fink: Climate Crisis Will Reshape Finance," February 24, 2020, https://www.nytimes.com/2020/01/14/business /dealbook/larry-fink-blackrock-climate-change.html.

7. B. McKibben, "Citing Climate Change, BlackRock Will Start Moving Away from Fossil Fuels," *New Yorker*, January 16, 2020, https://www.newyorker.com/news /daily-comment/citing-climate-change-blackrock-will-start-moving-away-from -fossil-fuels?verso=true. This follows calls for more environmental, social, and governance standards, or ESG, of the three biggest asset managers in the world, BlackRock, Vanguard, and State Street: *The Conversation*, "Three Financial Firms Could Change the Direction of the Climate Crisis—and Few People Have Any Idea," February 24, 2020, https://theconversation.com/three-financial-firms-could-change -the-direction-of-the-climate-crisis-and-few-people-have-any-idea-131869.

8. N. Bullard, "BlackRock's New Morality Marks the End for Coal," *Bloomberg*, January 17, 2020, https://www.bloomberg.com/opinion/articles/2020-01-17/black rock-s-climate-conscious-tidal-wave-breaks-on-coal.

9. The Rabbi did say elsewhere, in a later discussion, that there "has to be room for business" in climate solutions because "the banks aren't going anywhere."

10. N. Smith, "Nature as Accumulation Strategy," *Socialist Register* 43 (2007): 33.

11. Smith, "Nature as Accumulation Strategy," 33.

12. C. Katz, "Private Productions of Space and the 'Preservation' of Nature," in *Remaking Reality: Nature at the Millenium*, ed. B. Braun, and N. Castree (London: Routledge, 1998), 48.

13. Building on Moore, *Capitalism in the Web of Life*, I suggest that "green commodity frontiers" develop as capital seeks new opportunities to produce and market nature under, and within, conditions of ecological crisis. See also P. Bigger, J. Dempsey, A. P. Asiyanbi, K. Kay, R. Lave, B. Mansfield, T. Osborne, et al., "Reflecting on Neoliberal Natures: An Exchange," *Environment and Planning E: Nature and Space* 1, nos. 1–2 (2018): 25–75; K. McAfee and E. N. Shapiro, "Payments for Ecosystem Services in Mexico: Nature, Neoliberalism, Social Movements, and the State," *Annals of the Association of American Geographers* 100, no. 3 (2010): 579–99.

14. J. Dempsey, *Enterprising Nature: Economics, Markets, and Finance in Global Biodiversity Politics* (West Sussex, UK: John Wiley & Sons, 2016). See also S. Sullivan, "Banking Nature? The Spectacular Financialisation of Environmental Conservation," *Antipode* 45, no. 1 (2013): 198–217.

15. Klein, *The Shock Doctrine*. See also P. Le Billon, "Crisis Conservation and Green Extraction: Biodiversity Offsets as Spaces of Double Exception," *Journal of Political Ecology*, 28, no. 1: 854–88.

16. N. Klein, "How Power Profits from Disaster," *Guardian*, July 6, 2017, https:// www.theguardian.com/us-news/2017/jul/06/naomi-klein-how-power-profits -from-disaster.

17. M. Solis, "Coronavirus Is the Perfect Disaster for 'Disaster Capitalism,'" *Vice*, March 13, 2020, https://www.vice.com/en_uk/article/5dmqyk/naomi-klein-inter view-on-coronavirus-and-disaster-capitalism-shock-doctrine.

18. Sullivan, "Banking Nature?"

19. Corson, MacDonald, and Neimark, "Grabbing 'Green.'"

20. S. Ouma, L. Johnson, and P. Bigger, "Rethinking the Financialization of 'Nature,'" *Environment and Planning A: Economy and Space* 50, no. 3 (2018): 500–11; S. Bracking, "Financialization and the Environmental Frontier," in *The Routledge International Handbook of Financialization*, ed. D. Mertens, N. van der Zwan, and P. Mader (Abingdon, Oxon, UK: Routledge, 2020), 213–23.

21. Moore, "The Capitalocene Part II"; see also J. W. Moore, "Nature, Geopower, and Capitalogenic Appropriation," jasonwmoore.wordpress.com, November 8, 2016, https://jasonwmoore.wordpress.com/2016/11/08/nature-geopower-capitalogenic -appropriation/.

22. C. Parenti, "Environment-Making in the Capitalocene," in *Anthropocene or Capitalocene? Nature, History, and the Crisis of Capitalism*, ed. J. Moore (Oakland, Calif: PM Press, 2016), 166–83.

23. M. Robertson, "Measurement and Alienation: Making a World of Ecosystem Services," *Transactions of the Institute of British Geographers* 37, no. 3 (2012), 386–401, 3.

24. A. L. Tsing, "Inside the Economy of Appearances," *Public Culture* 12, no. 1 (2000): 115–44.

25. See this also in S. Milne and B. Adams, "Market Masquerades: Uncovering the Politics of Community-Level Payments for Environmental Services in Cambodia," *Development and Change* 43, no. 1 (2012): 133–58.

26. Some parts of this text first appeared in Neimark, "Biofuel Imaginaries."

27. M. Arsel and B. Büscher, "Nature™ Inc.: Changes and Continuities in Neoliberal Conservation and Market-Based Environmental Policy," *Development and Change* 43, no. 1 (2012): 53–78; B. Büscher, "Payments for Ecosystem Services as Neoliberal Conservation: (Reinterpreting) Evidence from the Maloti-Drakensberg, South Africa," *Conservation and Society* 10, no. 1 (2012): 29–41.

28. D. M. Lansing, "Understanding Smallholder Participation in Payments for Ecosystem Services: The Case of Costa Rica," *Human Ecology* 45, no. 1 (2017): 77–87.

29. B. Büscher, S. Sullivan, K. Neves, J. Igoe, and D. Brockington, "Towards a Synthesized Critique of Neoliberal Biodiversity Conservation," *Capitalism Nature Socialism* 23, no. 2 (2012): 4–30.

30. D. MacKenzie, "Is Economics Performative? Option Theory and the Construction of Derivatives Markets," *Journal of the History of Economic Thought* 28, no. 1 (2006): 29–55, 54, emphasis in original. See also P. Wes and J. Carrier, "Ecotourism and Authenticity: Getting Away from It All?" *Current Anthropology* 45, no. 4 (2004): 483–98.

31. Neimark, "Biofuel Imaginaries"; J. Fairhead and M. Leach, *Misreading the African Landscape: Society and Ecology in a Forest-Savanna Mosaic* (Cambridge: Cambridge University Press, 1996), 15. See also J. Igoe, *The Nature of the Spectacle: On Images, Money, and Conserving Capitalism* (Tucson: University of Arizona Press, 2017).

32. R. Peet and M. Watts, *Liberation Ecologies: Environment, Development and Social Movements* (London: Routledge, 1996), 37; T. M. Li, *The Will to Improve: Governmentality, Development, and the Practice of Politics* (Durham, N.C.: Duke University Press, 2007).

33. T. Forsyth, *Critical Political Ecology: The Politics of Environmental Science* (Abingdon, Oxon, UK: Routledge, 2004), 158.

34. N. R. Horning, "Strong Support for Weak Performance: Donor Competition in Madagascar," *African Affairs* 107, no. 428 (2008): 405–31.

35. J. W. Moore, "Sugar and the Expansion of the Early Modern World-Economy: Commodity Frontiers, Ecological Transformation, and Industrialization," *Review (Fernand Braudel Center)* (2000): 409–33, 409.

Chapter 2

1. According to Genese Sodikoff, "Contemporary Western imagery of Madagascar as a wounded, dying island—an image frequently invoked in conservation and ecotourism appeals—came into being through the imperialist encounter.'" Sodikoff, *Forest and Labor in Madagascar*, 37.

2. A. L. Zhu and B. Klein, "The Rise of Flexible Extraction: Boom-Chasing and Subject-Making in Northern Madagascar," *Geoforum* (2022), https://doi.org/10.1016/j.geoforum.2022.06.005.

3. A. Walsh, *Made in Madagascar: Sapphires, Ecotourism, and the Global Bazaar* (Toronto: University of Toronto Press, 2012).

4. W. Rodney, *How Europe Underdeveloped Africa* (London: Verso Trade, 2018). For a critique of the term "resource curse," see M. Watts, "Resource Curse? Governmentality, Oil and Power in the Niger Delta, Nigeria," *Geopolitics* 9, no. 1 (2004): 50–80.

5. K. J. Bakker, *An Uncooperative Commodity: Privatizing Water in England and Wales* (Oxford: Oxford Geographical and Environmental Studies, 2003).

6. For an excellent discussion on this, see I. R. Scales, "Paying for Nature: What Every Conservationist Should Know About Political Economy," *Oryx* 49, no. 2 (2015): 226–31. See also Bakker, *An Uncooperative Commodity*, 224.

7. Madagascar is rather exceptional—from its earliest days identified as an island of rare beauty that was continually facing threats to its survival at the hands of humankind. One only needs to look at the 2008 biofuel land deal that was set to lease 1.3 million ha to the South Korean multinational Daewoo Co. and subsequent coup d'état to see this tension of foreign and domestic resource politics playing out; see Neimark, "Biofuel Imaginaries."

8. J. C. Scott, "Seeing Like a State," in *Seeing Like a State: How Certain Schemes to Improve the Human Condition Have Failed* (New Haven, Conn.: Yale University

Press, 1998), 24. See also K. Paprocki, "The Climate Change of Your Desires: Climate Migration and Imaginaries of Urban and Rural Climate Futures," *Environment and Planning D: Society and Space* 38, no. 2 (2020): 248–66.

9. Moore, *Capitalism in the Web of Life*, 38; Moore, "Sugar and the Expansion of the Early Modern World-Economy"; B. J. Marley, "The Coal Crisis in Appalachia: Agrarian Transformation, Commodity Frontiers and the Geographies of Capital," *Journal of Agrarian Change* 16, no. 2 (2016): 225–54.

10. For excellent analyses that I draw on to a large extent concerning the history of market conservation and development, and its shortcomings, see Corson, *Corridors of Power*; Sodikoff, *Forest and Labor in Madagascar*; C. A. Kull, *Isle of Fire: The Political Ecology of Landscape Burning in Madagascar* (Chicago: University of Chicago Press, 2004); J. Pollini, "Slash-and-Burn Cultivation and Deforestation in the Malagasy Rain Forests: Representations and Results" (PhD diss., Cornell University, 2007).

11. P. Verin, "Deux étranges statues en chloritoschiste de Madagascar," *Publications de la Société Française D'histoire des Outre-mers* 5, no. 1 (1981): 155–60.

12. P. M. Larson, *History and Memory in the Age of Enslavement: Becoming Merina in Highland Madagascar, 1770–1822* (Portsmouth, N.H.: Heinemann, 2000).

13. Kull, *Isle of Fire*.

14. World Bank country data accessed at https://data.worldbank.org/?locations =MG.

15. J. Klein, B. Réau, and M. Edwards, "Zebu Landscapes: Conservation and Cattle in Madagascar," in *Greening the Great Red Island: Madagascar in Nature and Culture*, ed. J. Kaufmann (Africa Institute of South Africa, 2008), 157–78.

16. World Bank, "Madagascar: Balancing Conservation and Exploitation of Fisheries Resources," June 8, 2020, https://www.worldbank.org/en/news/feature/2020 /06/08/madagascar-balancing-conservation-and-exploitation-of-fisheries -resources.

17. J. Hassell, M. Roser, E. Ortiz-Ospina, and P. Arriagada, "Global Extreme Poverty," OurWorldInData, n.d., https://ourworldindata.org/poverty.

18. There are four principal types of rice growing: The first is irrigated "patty" rice, which is the majority of rice in terms of overall cultivated area. The second is rice grown on either the lowland slopes or valleys. The other two include upland hills (*tanety*) and in swiddens (*tavy*) or shifting cultivation.

19. S. Osterhoudt, S. S. Galvin, D. J. Graef, A. K. Saxena, and M. R. Dove, "Chains of Meaning: Crops, Commodities, and the 'In-Between' Spaces of Trade," *World Development* 135 (2020): 105070.

20. Osterhoudt et al., "Chains of Meaning."

21. GRiSP (Global Rice Science Partnership), *Rice Almanac*, 4th ed. (Los Baños, Philippines: International Rice Research Institute, 2013), 283, http://books.irri .org/9789712203008_content.pdf.

22. C. Moser, C. Barrett, and B. Minten, "Spatial Integration at Multiple Scales: Rice Markets in Madagascar," *Agricultural Economics* 40, no. 3 (2009): 281–94.

23. Moser, Barrett, and Minten, "Spatial Integration at Multiple Scales," 281–94.

24. E. Styger, J. Rakotoarimanana, R. Rabevohitra, and E. Fernandes, "Indigenous Fruit Trees of Madagascar: Potential Components of Agroforestry Systems to Improve Human Nutrition and Restore Biological Diversity," *Agroforestry Systems* 46, no. 3 (1999): 289–310.

25. Pollini, "Slash-and-Burn Cultivation and Deforestation."

26. P. W. Hanson, "Engaging Green Governmentality Through Ritual: The Case of Madagascar's Ranomafana National Park," *Études Océan Indien* 42–43 (2009): 85–113.

27. G. Althabe, *Oppression et liberation dans l'imaginaire, les communautés villageoises le la côte orientale de Madagascar* (Paris: F. Maspero, 1969).

28. Kull, *Isle of Fire.*

29. Kull, *Isle of Fire.*

30. Styger et al., "Indigenous Fruit Trees of Madagascar," 289–310.

31. L. Jarosz, "Defining and Explaining Tropical Deforestation: Shifting Cultivation and Population Growth in Colonial Madagascar (1896–1940)," *Economic Geography* 69, no. 4 (1993): 366–79.

32. I. R. Scales, "The Future of Conservation and Development in Madagascar: Time for a New Paradigm?" *Madagascar Conservation and Development* 9, no. 1 (2014): 5–12.

33. Original estimates of 90 percent of the island being deforested have been heavily questioned due to lack of precise data and assumptions of previous size and composition of forests. See Scales, "Future of Conservation"; G. Vieilledent, C. Grinand, F. A. Rakotomalala, R. Ranaivosoa, J-R Rakotoarijaona, T. Allnutt, and F. Achard, "Combining Global Tree Cover Loss Data with Historical National Forest Cover Maps to Look at Six Decades of Deforestation and Forest Fragmentation in Madagascar," *Biological Conservation* 222 (2018): 189–97.

34. Sodikoff, *Forest and Labor in Madagascar.*

35. Malagasy smallholders are usually defined as a category of rural farmer who holds an average land holding of roughly 1.5 hectares. Yet there is significant heterogeneity between and within smallholder groups across the county, and many farming activities, including subsistence and cash crop production, animal husbandry, and forestry and artisanal fisheries.

36. Verin, "Deux étranges statues."

37. P. M. Allen and M. Covell, *Historical Dictionary of Madagascar*, Historical Dictionaries of Africa, no. 98 (Lanham, Md.: Scarecrow Press, 2005).

38. G. Campbell, *An Economic History of Imperial Madagascar, 1750–1895: The Rise and Fall of an Island Empire* (Cambridge: Cambridge University Press, 2005).

39. R. E. Dewar and A. F. Richard, "Madagascar: A History of Arrivals, What Happened, and Will Happen Next," *Annual Review of Anthropology* 41 (2012): 495–517.

40. J. Hooper, *Feeding Globalization: Madagascar and the Provisioning Trade, 1600–1800* (Athens: Ohio University Press, 2017).

41. Allen and Covell, *Historical Dictionary of Madagascar.*
42. E. A. Alpers, "Recollecting Africa: Diasporic Memory in the Indian Ocean World," *African Studies Review* 43, no. 1 (2000): 83–99.
43. G. Campbell, "The Structure of Trade in Madagascar, 1750–1810," *International Journal of African Historical Studies* 26, no. 1 (1993): 111–48.
44. Allen and Covell, *Historical Dictionary of Madagascar.*
45. M. Brown, "Madagascar: Island of the Ancestors," *Anthropology Today* 3, no. 1 (1987): 14–17.
46. R. K. Kent, *Early Kingdoms in Madagascar, 1500–1700* (New York: Holt, Rinehart and Winston, 1970).
47. J. Cole, *Forget Colonialism? Sacrifice and the Art of Memory in Madagascar* (Berkeley: University of California Press, 2001).
48. Campbell, "Structure of Trade in Madagascar," 111–48, 127.
49. Campbell, "Structure of Trade in Madagascar," 127.
50. G. Campbell, ed., *Bondage and the Environment in the Indian Ocean World* (Cham: Springer, 2018), 92.
51. Scales, "Future of Conservation."
52. S. Evers, G. Campbell, and M. Lambek, "Land Competition and Human-Environment Relations in Madagascar," in *Contest for Land in Madagascar: Environment, Ancestors and Development*, ed. S. Evers, G. Campbell, and M. Lambek (Leiden: Brill, 2013), 1–20; see also D. Henkels, "A Close Up of Malagasy Environmental Law," *Vermont Journal of Environmental Law* 3, no. 47 (2001).
53. R. Callet, *Tantaran'ny Andriana eto Madagascar* (Antananarivo: Académie Malgache, 1974). See also M. Hufty and F. Muttenzer, "Devoted Friends: The Implementation of the Convention on Biological Diversity in Madagascar," in *Governing Global Biodiversity*, ed. P. Le Prestre (Aldershot: Ashgate, 2002); S. Olson, "The Robe of the Ancestors: Forests in the History of Madagascar," *Journal of Forest History* 28, no. 4 (1984): 174–86, 178.
54. P. Montagne and B. Ramamonjisoa, "Politiques forestières à Madagascar entre répression et autonomie des acteurs," *Économie Rurale: Agricultures, Alimentations, Territoires* 4–5, nos. 294–95 (2006): 9–26.
55. G. Campbell, personal communication, 2020.
56. G. Campbell, personal communication, 2021.
57. Verin, "Deux étranges statues"; G. Campbell, "Forest Depletion in Imperial Madagascar, c. 1790–1861," in *Contest for Land in Madagascar: Environment, Ancestors and Development*, ed. S. Evers, G. Campbell, and M. Lambek (Leiden: Brill, 2013), 31, 63–95.
58. Evers, Campbell, and Lambek, "Land Competition."
59. Also known as the "Bloody Mary of Madagascar"; see A. Chernock, "Queen Victoria and the 'Bloody Mary of Madagascar,'" *Victorian Studies* 55, no. 3 (2013): 425–49.
60. S. Randrianja and S. Ellis, *Madagascar: A Short History* (London: C. Hurst, 2009). See also J. P. Jones, O. S. Rakotonarivo, and J. H. Razafimanahaka, "Forest

Conservation in Madagascar: Past, Present, and Future," in *The New Natural History of Madagascar*, ed. S. Goodman (Princeton, N.J.: Princeton University Press, 2021).

61. Corson, *Corridors of Power*, 38; see also Campbell, *An Economic History*; Evers, Campbell, and Lambek, "Land Competition."

62. Evers, Campbell, and Lambek, "Land Competition"; A. Keck, N. P. Sharma, and G. Feder, *Population Growth, Shifting Cultivation, and Unsustainable Agricultural Development: A Case Study in Madagascar* (Washington, D.C.: World Bank Publications, 1994).

63. In reality, state control was basically confined to the central highlands until the mid-nineteenth century. For more on this, see G. J. Harper, M. K. Steininger, C. J. Tucker, D. Juhn, and F. Hawkins, "Fifty Years of Deforestation and Forest Fragmentation in Madagascar," *Environmental Conservation* 34, no. 4 (2007): 325–33. See also Jones, Rakotonarivo, and Razafimanahaka, "Forest Conservation in Madagascar."

64. Cole, *Forget Colonialism?*

65. S. Goedefroit and J. Lombard, *Andolo: L'art funéraire sakalava à Madagascar* (Paris: IRD, 2007); Evers, Campbell, and Lambek, "Land Competition."

66. Cole, *Forget Colonialism?*, 42.

67. Feeley-Harnik, *Green Estate*, 391; see also L. A. Sharp, *The Sacrificed Generation* (Berkeley: University of California Press, 2002).

68. Campbell, "Forest Depletion in Imperial Madagascar," 63–95.

69. Jarosz, "Defining and Explaining Tropical Deforestation."

70. Jones, Rakotonarivo, and Razafimanahaka, "Forest Conservation in Madagascar"; Corson, *Corridors of Power*.

71. L. Lavauden, "Histoire de la législation et de l'administration forestière à Madagascar," *Revue des Eaux et Forêts* 72 (1934): 949–60; Pollini, "Slash-and-Burn Cultivation."

72. Lavauden, "Histoire de la législation," as cited in Pollini, "Slash-and-Burn Cultivation," 45.

73. Corson, *Corridors of Power*.

74. Kull, *Isle of Fire*; Sodikoff, *Forest and Labor in Madagascar*.

75. Evers, Campbell, and Lambek, "Land Competition."

76. For more on colonial scientific forestry, see P. Vandergeest and N. L. Peluso, "Empires of Forestry: Professional Forestry and State Power in Southeast Asia, Part 2," *Environment and History* 12, no. 4 (2006): 359–93; P. Montagne and A. Bertrand, *Histoire des politiques forestières au Niger, au Mali et à Madagascar* (Paris: L'Harmattan, 2006); Kull, *Isle of Fire*.

77. G. Ramanantsoavina, *Histoire de la politique forestière à Madagascar* (Antananarivo: DGEF, 1963), 831–52, 46.

78. Pollini, "Slash-and-Burn Cultivation"; Ramanantsoavina, *Histoire de la politique forestière*.

79. Ramanantsoavina, *Histoire de la politique forestière*.

80. T. Anderson, "Solving Madagascar: Science, Illustrations, and the Normalizing of Fauna of Nineteenth Century Madagascar," in *Contest for Land in Madagascar: Environment, Ancestors and Development,* ed. S. Evers, G. Campbell, and M. Lambek (Leiden: Brill, 2013), 97–118.

81. Anderson, "Solving Madagascar."

82. F. Andriamialisoa and O. Langrand, "The History of Zoological Exploration of Madagascar," in *The Natural History of Madagascar,* ed. S. Goodman and J. Benstead (Chicago: University of Chicago Press, 2003), 1–15. See also Feeley-Harnik, *Green Estate*; Sodikoff, *Forest and Labor in Madagascar*; Anderson, "Solving Madagascar."

83. For an excellent summary on the legacy of colonial botanical collections, see L. Schiebinger, *Plants and Empire: Colonial Bioprospecting in the Atlantic World* (Harvard University Press, 2007).

84. Andriamialisoa and Langrand, "History of Zoological Exploration."

85. Pollini, "Slash-and-Burn Cultivation"; Feeley-Harnik, *Green Estate*; Corson, *Corridors of Power.*

86 Corson, *Corridors of Power.*

87 Anderson, "Solving Madagascar."

88. *L'Express de Madagascar,* "Un jardin d'essais cultural a Namisana," December 28, 2016, https://lexpress.mg/28/12/2016/un-jardin-dessais-culturaux-a-nanisana/.

89. Keck, Sharma, and Feder, *Population Growth*; Sodikoff, *Forest and Labor in Madagascar.*

90. Jarosz, "Defining and Explaining."

91. Jarosz, "Defining and Explaining," 370; Campbell, *An Economic History.*

92. S. R. Osterhoudt, *Vanilla Landscapes: Meaning, Memory, and the Cultivation of Place in Madagascar* (New York: New York Botanical Garden, 2017); see also Jarosz, "Defining and Explaining."

93. Corson, *Corridors of Power,* 38; see also Jones, Rakotonarivo, and Razafimanahaka, "Forest Conservation in Madagascar."

94. Jarosz, "Defining and Explaining."

95. J. Harper, "Memories of Ancestry in the Forests of Madagascar," in *Landscape, Memory and History,* ed. P. J. Stewart and A. Strathern (London: Pluto Press, 2003), 89–107.

96. Althabe, *Oppression et liberation dans l'imaginaire*; Cole, *Forget Colonialism?*; Kull, *Isle of Fire.*

97. Cole, *Forget Colonialism?,* 40.

98. Cole, *Forget Colonialism?,* 41.

99. Sodikoff, *Forest and Labor in Madagascar,* 414.

100. Sodikoff, *Forest and Labor in Madagascar,* 51. Called the Service de la Main-d 'Oeuvre des Travaux d'Interêt Général (SMOTIG).

101. Jarosz, "Defining and Explaining," 370.

102. Jarosz, "Defining and Explaining," 357; J. Hornac, "Le deboisement et la politique forestiére à Madagascar: Mémoire de stage," in *Mémoires de l'ecole colo-*

niale ENFOM (Aix-en-Provence: Archives Nationales de France, Archives D'Outre Mer).

103. G. Ramanantsoavina, and A. Rakotomanampison, *Fertilisation des plantations industrielles de pins à Madagascar* (Antananarive: GERDAT-CTFT, 1973). See also Campbell, *An Economic History*;

104. Corson, *Corridors of Power*, 46.

105. Harper, "Memories of Ancestry."

106. Kull, *Isle of Fire.*

107. Harper, "Memories of Ancestry."

108. Lavauden, "Histoire de la législation."

109. Montagne and Bertrand, *Histoire des politiques forestières.*

110. Corson, *Corridors of Power*, 46.

111. R. R. Marcus and A. M. Ratsimbaharison, "Political Parties in Madagascar: Neopatrimonial Tools or Democratic Instruments?" *Party Politics* 11, no. 4 (2005): 495–512.

112. A. M. Ratsimbaharison, "The Obstacles and Challenges to Democratic Consolidation in Madagascar (1992–2009)," *International Journal of Political Science* 2, no. 2 (2016): 1–17.

113. Scales, "Future of Conservation."

114. IMF, *Republic of Madagascar: Poverty Reduction Strategy Paper Annual Progress Report* (Washington, D.C.: IMF, 2005).

115. H. Metz, *Indian Ocean: Five Island Countries* (Washington, D.C.: Federal Research Division, 1995).

116. D. Raik, "Forest Management in Madagascar: An Historical Overview," *Madagascar Conservation and Development* 2, no. 1 (2007).

117. R. R. Marcus, and C. A. Kull, "The Politics of Conservation in Madagascar," *African Studies Quarterly* 3, no. 2 (1999): 1–8; see also Jones, Rakotonarivo, and Razafimanahaka, "Forest Conservation in Madagascar."

118. Initially the NEAP set up a comprehensive system of protected areas, aiming to stem deforestation and biodiversity loss. At that time, only two national parks existed alongside a dozen or so natural reserves, representing less than 2 percent of the country. Many of these were essentially paper parks with few resources devoted to their maintenance. For a more comprehensive history of NEAP, see Freudenberger and the International Resources Group, *Paradise Lost?*

119. Corson, *Corridors of Power.*

120. A. Hewitt, "Madagascar," in *Structural Adjustment and the African Farmer*, ed. A. Duncan and J. Howell (London: Overseas Development Institute, 1992). See also Kull, *Isle of Fire.*

121. R. A. Schroeder, "Community, Forestry and Conditionality in the Gambia," *Africa* 69, no. 1 (1999): 1–22. See also R. A. Schroeder and R. P. Neumann, "Manifest Ecological Destinies: Local Rights and Global Environmental Agendas," *Antipode* 27, no. 4 (1995): 321–24.

122. R. Duffy, "Non-governmental Organisations and Governance States: The Impact of Transnational Environmental Management Networks in Madagascar," *Environmental Politics* 15, no. 5 (2006): 731–49.

123. Freudenberger and the International Resources Group, *Paradise Lost?*, 9; see also Horning, "Strong Support for Weak Performance."

124. R. Randriamanamisa and G. J. Stads, *Madagascar: Recent Developments in Public Agricultural Research*, Agricultural Science and Technology Indicators (Rome: Food and Agriculture Organization of the United Nations, 2010).

125. Freudenberger and the International Resources Group, *Paradise Lost?*

126. Integrated Conservation and Development policy was followed by the original eight Millennium Development Goals and the subsequent transition to the now seventeen Sustainable Development Goals.

127. E. Keller, "The Banana Plant and the Moon: Conservation and the Malagasy Ethos of Life in Masoala, Madagascar," *American Ethnologist* 35, no. 4 (2008): 650–64.

128. World Bank, United Nations Development Program, *Coopération Suisse, Coopération Française*, the Food and Agriculture Organization of the UN (FAO) and the European Union (EU). See Kull, *Isle of Fire*.

129. Environmental NGOs include the World Wildlife Fund for Nature (WWF) and Conservation International (CI).

130. Research universities include Duke, Cornell, and Stony Brook. A. Bertrand and M. Sourdat, *Feux et Déforestation á Madagascar: Revues Bibliographiques* (Antananarivo: CIRAD/ORSTOM/CITE, 1998); Kull, *Isle of Fire*.

131. This was awarded through an International Development Association Grant (IDA).

132. Marcus and Kull, "The Politics of Conservation in Madagascar."

133. For more on GELOSE, see Pollini, "Slash-and-Burn Cultivation."

134. For more on *dina*, community relations, and commodity production, see B. I. Klein, "Dina, Domination, and Resistance: Indigenous Institutions, Local Politics, and Resource Governance in Madagascar," *Journal of Peasant Studies* (2023): 1–30.

135. UNCTAD, *Foreign Direct Investment in LDCs: Lessons Learned from the Decade 2001–2010 and the Way Forward* (New York: United Nations, 2011).

136. P. Burnod, M. Gingembre, and R. Andrianirina Ratsialonana, "Competition Over Authority and Access: International Land Deals in Madagascar," *Development and change* 44, no. 2 (2013): 357–79.

137. A. Teyssier, R. Andrianirina-Ratsialonana, R. Razafindralambo, Y. Razafindrakoto, "Decentralization of Land Management in Madagascar: Process, Innovations, and Observation of the First Outcomes." Paper presented at the Annual World Bank Conference on Land Administration, Washington, D.C., 2008.

138. R. Andrianirina-Ratsialonana, L. Ramarojohn, P. Burnod, and A. Teyssier, *After Daewoo? Current Status and Perspectives of Large-Scale Land Acquisition in Madagascar* (2011), https://agritrop.cirad.fr/560265/1/document_560265.pdf.

139. Neimark, "Biofuel Imaginaries."

140. Program of Governorship of Mineral Resources (PRGM). See Report No. ICR2208 (IDA-37540 IDA-37541 IDA-3754A).

141. B. McCourt, *BGS International Activities* (2010), http://nora.nerc.ac.uk/id/eprint /9191/.

142. B. Sarrasin, "The Mining Industry and the Regulatory Framework in Madagascar: Some Developmental and Environmental Issues," *Journal of Cleaner Production* 14, nos. 3–4 (2006): 388–96.

143. F. Seymour and J. Busch, *Why Forests? Why Now? The Science, Economics, and Politics of Tropical Forests and Climate Change* (Washington, D.C.: Center for Global Development, 2016).

144. Jones, Rakotonarivo, and Razafimanahaka, "Forest Conservation in Madagascar."

145. "Forest Carbon Partnerships, Readiness Package (R-Package) for Reducing Emissions from Deforestation and Forest Degradation in Madagascar," 2017, https://www.forestcarbonpartnership.org/system/files/documents/Madagascar -TAP%20R-Package%20Review-10%20Septj.pdf.

146. B. Ferguson, "REDD Comes into Fashion in Madagascar," *Madagascar Conservation & Development* 4, no. 2 (2009), https://www.ajol.info/index.php/mcd /article/view/48654.

147. C. Grinand, F. Rakotomalala, V. Gond, R. Vaudry, M. Bernoux, and G. Vieilledent, "Estimating Deforestation in Tropical Humid and Dry Forests in Madagascar from 2000 to 2010 Using Multi-Date Landsat Satellite Images and the Random Forests Classifier," *Remote Sensing of Environment* 139 (2013): 68–80, https://doi.org/10.1016/j.rse.2013.07.008.

148. J. M. M. Yancho, T. G. Jones, S. R. Gandhi, C. Ferster, A. Lin, and L. Glass, "The Google Earth Engine Mangrove Mapping Methodology (GEEMMM)," *Remote Sensing* 12, no. 22 (2020): 3758.

149. L. Brimont, D. Ezzine-de-Blas, A. Karsenty, and A. Toulon, "Achieving Conservation and Equity Amidst Extreme Poverty and Climate Risk: The Makira REDD+ Project in Madagascar," *Forests* 6, no. 3 (2015): 748–68.

150. Jones, Rakotonarivo, and Razafimanahaka, "Forest Conservation in Madagascar."

151. L. Gifford, "'You Can't Value What You Can't Measure': A Critical Look at Forest Carbon Accounting," *Climatic Change* 161, no. 2 (2020): 291–306.

152. R. Duffy, "Global Environmental Governance and the Politics of Ecotourism in Madagascar," *Journal of Ecotourism* 5, nos. 1–2 (2006): 128–44.

153. K. Marx, *Capital. A Critique of Political Economy*, vol. 1 (London: Penguin Books, 1976).

154. B. Büscher, "Letters of Gold: Enabling Primitive Accumulation Through Neoliberal Conservation," *Human Geography* 2, no. 3 (2009): 91–94, 91.

Chapter 3

1. BBC, "Coronavirus: Caution Urged over Madagascar's 'Herbal Cure,'" April 22, 2020, https://www.bbc.co.uk/news/world-africa-52374250.

Some sections of this chapter are reprinted from Neimark, "Green Grabbing"; Neimark and Wilson, "Re-mining the Collections"; B. D. Neimark and S. Vermeylen, "A Human Right to Science? Precarious Labor and Basic Rights in Science and Bioprospecting," *Annals of the American Association of Geographers* 107, no. 1 (2017): 167–82.

2. For an excellent review of Covid-Organics and a deep-dive into the history of Malagasy postcolonial herbal medicine, see C. Adams, "The DJ and the Miracle Cure: The Perils of Postcolonial Medicine in Madagascar," *Drift*, October 21, 2020, https://www.thedriftmag.com/the-dj-and-the-miracle-cure/.

3. Not coincidentally for the flamboyant, self-promoting president, these are the colors of his political party, Young Malagasies Determined (Tanora malaGasy Vonona, or TGV).

4. A. Vaughan, "No Evidence 'Madagascar Cure' for COVID-19 Works, says WHO," *New Scientist*, May 15, 2020, https://www.newscientist.com/article/2243669-no-evidence-madagascar-cure-for-COVID-19-works-says-who/#ixzz6Oh0NaQVZ.

5. BBC, "Coronavirus."

6. M. Perelman, "Exclusive: Madagascar's President Defends Controversial Homegrown COVID-19 Cure," *France 24*, December 5, 2020, https://www.france24.com/en/africa/20200512-exclusive-madagascar-s-president-defends-controversial-homegrown-COVID-19-cure.

7. Rajoelina said in response to the WHO's concerns: "The proof of the tonic's efficacy was in the 'healing' of 'our patients,'" calling it a "preventive and curative remedy." Reminding viewers that Madagascar has a long history of traditional medicine, Rajoelina pointed out that many pharmaceutical drugs authorized in the West have turned out to be harmful, such as the Mediator weight loss drug in France.

8. Perelman, "Exclusive." The Malagasy scientists being referred to here were the researchers at the Malagasy National Institute for Applied Pharmaceuticals (IMRA) and its famous founder, Albert Rakoto Ratsimamanga. Malagasy scientists were indeed at the heart of the research. I was familiar with IMRA, the institution that made Covid-Organics, having visited the laboratory in the past. By no means was IMRA a "chop-shop." Rather; it was an internationally respected laboratory maintaining long-standing relationships with global pharmaceutical firms and has a history of drug discovery research from nature.

9. See my extensive study of the remarkable discovery and periwinkle's commodity chain in Neimark, "Green Grabbing."

10. See Adams, "DJ and the Miracle Cure."

11. Sometimes referred to as a circular economy.

12. European Commission, "Report on Innovating for Sustainable Growth: A Bioeconomy for Europe," #A7–0201/2013 (Brussels, 2013), https://www.europarl.europa.eu/doceo/document/A-7-2013-0201_EN.html.

13. European Commission, "New Perspectives on the Knowledge-Based Bio-Economy," Conference Report (Brussels: DG-Research, 2005), 2; quoted in

K. Birch and D. Tyfield, "Theorizing the Bioeconomy: Biovalue, Biocapital, Bio-economics or . . . What?" *Science, Technology & Human Values* 38, no. 3 (2013): 299–327, 303.

14. K. Birch, L. Levidow, and T. Papaioannou, "Sustainable Capital? The Neolib-eralization of Nature and Knowledge in the European 'Knowledge-Based Bio-economy,'" *Sustainability* 2, no. 9 (2010): 2898–918, 2899.

15. Anon #3–1C.

16. S. Benabou, "Making Up for Lost Nature? A Critical Review of the Interna-tional Development of Voluntary Biodiversity Offsets," *Environment and Society* 5, no. 1 (2014): 103–23; C. Seagle, "Inverting the Impacts: Mining, Conservation and Sustainability Claims near the Rio Tinto/QMM Ilmenite Mine in Southeast Madagascar," *Journal of Peasant Studies* 39, no. 2 (2012): 447–77.

17. D. D. Soejarto, H. H. S. Fong, G. T. Tan, H. J. Zhang, C. Y. Ma, S. G. Franzblau, and G. R. Dietzman, "Ethnobotany/Ethnopharmacology and Mass Bioprospect-ing: Issues on Intellectual Property and Benefit-Sharing," *Journal of Ethnophar-macology* 100, nos. 1–2 (2005): 15–22.

18. R. Wynberg and S. Laird, "Bioprospecting: Tracking the Policy Debate," *Envi-ronment: Science and Policy for Sustainable Development* 49, no. 10 (2007): 20–32.

19. K. Ten Kate and S. A. Laird, "Biodiversity and Business: Coming to Terms with the 'Grand Bargain,'" *International Affairs* 76, no. 2 (2000): 241–64.

20. In the late 1970s, a company called Smith Kline & French developed an ulcer medicine called Tagamet, thought to be the first blockbuster drug and provid-ing the company with a windfall of $1 billion and the Nobel Prize.

21. K. Ten Kate and S. A. Laird, *The Commercial Use of Biodiversity: Access to Genetic Resources and Benefit-Sharing* (Milton, UK: Routledge, 2019). See also J. S. Miller, "The Discovery of Medicines from Plants: A Current Biological Per-spective," *Economic Botany* 65, no. 4 (2011): 396–407.

22. T. Eisner, "Chemical Prospecting: A Global Imperative," *Proceedings of the Amer-ican Philosophical Society* 138, no. 3 (1994): 385–93. See also M. J. Balick, "Eth-nobotany and the Identification of Therapeutic Agents from the Rainforest," in *Ciba Foundation Symposium 154: Bioactive Compounds from Plants* (Chiches-ter, UK: John Wiley & Sons, 2007), 22–39.

23. C. Hayden, *When Nature Goes Public: The Making and Unmaking of Biopros-pecting in Mexico*, vol. 1 (Princeton: Princeton University Press, 2003), 57.

24. Hayden, *When Nature Goes Public*, 58–59.

25. B. D. Neimark and R. A. Schroeder, "Hotspot Discourse in Africa: Making Space for Bioprospecting in Madagascar," *African Geographical Review* 28, no. 1 (2009): 43–69.

26. C. Macilwain, "When Rhetoric Hits Reality in Debate on Bioprospecting," *Nature* 392, no. 6676 (1998): 535.

27. W. Reid, *Biodiversity Prospecting: Using Genetic Resources for Sustainable Devel-opment* (Washington D.C.: WRI, 1993).

28. Nagoya Protocol, *Access to Genetic Resources and the Fair and Equitable Sharing of Benefits Arising from their Utilization (ABS)*, was adopted on October, 29, 2010, in Nagoya, Japan, and entered into force on October 12, 2014, updating the CBD and, through a predetermined agreement, providing access to a country's biogenetic resources in return for commercialized benefits. https://www.cbd.int/abs/about/.

29. B. Parry, *Trading the Genome: Investigating the Commodification of Bio-information* (New York: Columbia University Press, 2004), 374. For more on the politics of bioprospecting globally, see V. Shiva, *Biopiracy: The Plunder of Knowledge and Nature* (Boston: South End Press, 1997); H. Svarstad, "A Global Political Ecology of Bioprospecting," in *Political Ecology Across Spaces, Scales and Social Groups*, ed. S. Paulson and L. Gezon (New Brunswick, N.J.: Rutgers University Press, 2004), 239–56.

30. F. E. Koehn and G. T. Carter, "The Evolving Role of Natural Products in Drug Discovery," *Nature Reviews Drug Discovery* 4, no. 3 (2005): 206–20; Ten Kate and Laird, *The Commercial Use of Biodiversity*.

31. Reid, *Biodiversity Prospecting*, 2; A. Sittenfeld and A. Lovejoy, "Biodiversity Prospecting Frameworks: The INBio Experience in Costa Rica," in *Protection of Global Biodiversity: Converging Strategies*, ed. L. Guruswamy and J. McNeely (Durham, N.C.: Duke University Press, 1996), 223–44.

32. B. Neimark and L. Tilghman, "Bioprospecting a Biodiversity Hotspot: The Political Economy of Natural Products Drug Discovery for Conservation Goals in Madagascar," in *Conservation and Environmental Management in Madagascar*, ed. I. R. Scales (London: Routledge, 2014), 295–322.

33. Findings also came from the National Science Foundation and United States Agency for International Development (USAID). The Foreign Agriculture Service of the USDA eventually replaced USAID. By 1999, the ICBG had reported the collection of 11,000 samples and up to 200,000 different types of therapeutic screens and had located 260 active compounds, 60 of them reported as "novel." See J. Rosenthal, F. Katz, and A. Bull, *Microbial Diversity and Bioprospecting* (Washington, D.C.: ASM Press, 2004).

34. This term was traditionally was used to signify white Europeans during the colonial period, and is sometimes used by ethnic groups on the coasts to refer to people of the highland Merina group.

35. The short anecdote that follows was originally published in Neimark and Vermeylen, "A Human Right to Science?"

36. All names of research participants are pseudonyms. Malagasy village names have also been changed.

37. G. Campbell, "Madagascar and Mozambique in the Slave Trade of the Western Indian Ocean, 1800–1861," *Slavery and Abolition* 9, no. 3 (1988): 165–92.

38. J. Martinez-Alier, *The Environmentalism of the Poor: A Study of Ecological Conflicts and Valuation* (Northhampton, Mass.: Edward Elgar, 2002), 22.

39. S. A. Laird, *Biodiversity and Traditional Knowledge: Equitable Partnerships in Practice* (n.p.: Routledge, 2010).

40. B. Neimark, "Bioprospecting and Biopiracy," in *The International Encyclopedia of Geography: People, the Earth, Environment and Technology*, ed. D. Richardson, N. Castree, M. F. Goodchild, A. L. Kobayashi, W. Liu, and R. A. Marston (Chichester, UK: John Wiley & Sons, 2017).

41. See Shiva, *Biopiracy*. Important work on biopiracy has been done by the civil-society group ETC led by Pat Mooney: https://www.etcgroup.org/users/pat-mooney.

42. Anon #3–1D.

43. Hayden, *When Nature Goes Public*, 57.

44. Parry, *Trading the Genome*, 55.

45. Hayden, *When Nature Goes Public*.

46. D. Brockington and R. Duffy, eds., *Capitalism and Conservation* (Malden, Mass.: John Wiley & Sons, 2011).

47. Some of the history from this section is republished from Neimark and Tilghman, "Bioprospecting a Biodiversity Hotspot."

48. W. Sneader, *Drug Discovery: A History* (Hoboken, N.J.: John Wiley & Sons, 2005).

49. G. M. Cragg and D. J. Newman, "Biodiversity: A Continuing Source of Novel Drug Leads," *Pure and Applied Chemistry* 77, no. 1 (2005): 7–24.

50. Of particular importance was Avicenna's *Canon Medicinae*, published in 1025; for an extensive history, see Sneader, *Drug Discovery*.

51. L. Brockway, "Science and Colonial Expansion: The Role of the British Royal Botanic Gardens," *American Ethnologist* 6, no. 3 (1979): 449–65. For more on colonial botany, see L. Schiebinger and C. Swan, eds., *Colonial Botany: Science, Commerce, and Politics in the Early Modern World* (Philadelphia: University of Pennsylvania Press, 2007).

52. Sneader, *Drug Discovery*.

53. J. Drews, "Drug Discovery: A Historical Perspective," *Science* 287, no. 5460 (2000): 1960–64.

54. Cragg and Newman, "Biodiversity."

55. These contracts, which averaged $2.7 million, obligated the institutions to collect plant samples, voucher specimens, and associated botanical, ecological, and ethnobotanical information that could lead to new drug discoveries. The collection sites included tropical and subtropical locations in South and Central America, Africa, and Asia. Additionally, the NCI signed individual collaborative ventures with selective institutions and researchers in other countries. Following this drive for drugs from nature was another initiative spearheaded by the Natural Products Branch (NPB) of the NCI's Division of Cancer Treatment and Diagnosis, which began a massive program of collecting biological resources worldwide. Cragg and Newman, "Biodiversity." For more on this history, see G. Cragg and D. Newman, "Natural Product Drug Discovery in the Next Millennium," *Pharmaceutical Biology* 39 (2001): 8–17.

56. Macilwain, "When Rhetoric Hits Reality."

57. Koehn and Carter, "Evolving Role of Natural Products," 206–20.

58. S. Laird, *Access and Benefit-Sharing: Key Points for Policy-Makers, the Pharmaceutical Industry*, The ABS Capacity Development Initiative, GIZ (Cape Town: University of Cape Town and People and Plants International, 2015), 7.

59. Laird, *Access and Benefit-Sharing*, 7.

60. Another major trend in the industry comprises developing technologies of synthetic biology, biomimicry, and the emerging work on reengineering or editing of life-forms using models based on nature. The recent mapping of the first synthetic genome may actually revive some programs that use natural compounds as a model precursor for new drug discoveries.

61. For a more extensive discussion, see B. Neimark, *Industrial Heartlands of Nature: The Political Economy of Biological Prospecting in Madagascar* (New Brunswick, N.J.: Rutgers, 2009).

62. See Neimark, "Green Grabbing." Another example of a supply issue of a natural product includes the collection of the pacific yew (*Taxus brevifolia*) for the anticancer drug paclitaxel. Due to the scarcity of the woody shrub and overexploitation in the wild, a cultivation method of its closely related cousin (*Taxus baccata*) was soon devised, and enough bioactive compounds in the bark were made available to researchers for isolation of the synthetic compounds. The use of the "cousin" plant is an example of taxonomically guided bioprospecting.

63. D. Kingston, "Successful Drug discovery from Natural Products: Methods and Results." Paper read at the Proceedings of the 11th NAPRECA Symposium, at Antananarivo, Madagascar, 2006.

64. This was expressed by a number of informants.

65. Anon #3–1D.

66. Anon #3–1E.

67. To date, the Uruguay Round was most notable for the transformation of the General Agreement on Tariffs and Trade (GATT) to the WTO. The agreement included trade tariffs, barriers, and subsidies pertaining to textiles and clothing, agriculture, and tropical produce. Its subsequent agreement, TRIPS, began the discussion pertaining to IPRs and commercialization and paved the way for governments to approach the basic issues of copyright, patents, trademarks, industrial-trade secrets, and geographic origin.

68. The TRIPS agreement was not the only international framework that addressed intellectual property. Several agreements have attempted to recognize traditional access rights and "cultural" knowledge in more formal terms. Other examples include the 1988 International Conference of Belém, Brazil, and the second Code of Ethics of the International Society of Ethnobotany, codified in 1991 in Kunming, China. Subsequently, both agreements were codified under the CBD in 1992, requiring bioprospectors to obtain informed consent from all the parties involved in the exchange and share benefits with those who supplied the knowledge.

69. J. Miller, C. Birkinshaw, and M. Callmander, "The Madagascar International Cooperative Biodiversity Group (ICBG): Using Natural Products Research to Build Science Capacity," *Ethnobotany Research and Applications* 3 (2005): 283–86.

70. Anon #3–1E.

71. D. Robinson, *Biodiversity, Access and Benefit-Sharing: Global Case Studies* (Oxon, UK: Routledge, 2014), http://dx.doi.org/10.4324/9781315882819.
72. D. Kingston, "Bioprospecting for Biodiversity Conservation in Madagascar," REEIS, 2013, https://reeis.usda.gov/web/crisprojectpages/0215326-biodiversity -conservation-and-drug-discovery-in-madagascar.html.
73. Selections of this section taken from Neimark and Wilson, "Re-mining the Collections."
74. Anon #3–1F.
75. Anon #3–1G.
76. M. Ratsimbason, L. Ranarivelo, H. R. Juliani, and J. E. Simon, "Antiplasmodial Activity of Twenty Essential Oils from Malagasy Aromatic Plants," in *African Natural Plant Products: New Discoveries and Challenges in Chemistry and Quality*, ed. H. R. Juliani, J. E. Simon, and C. T. Ho (Washington, D.C.: American Chemical Society, 2009), 209–15; S. Cao and D. G. Kingston, "Biodiversity Conservation and Drug Discovery: Can They Be Combined? The Suriname and Madagascar Experiences," *Pharmaceutical Biology* 47, no. 8 (2009): 809–23; B. H. Rakotonjatovo, A. Rasolofoarimanga, H. Andriamanantoanina, L. Ranarivelo, J. Maharavo, L. Ramaroson, and M. Ratsimbason, "A Microfluorimetric Method to Screen Marine Products for Antimalarial Activity: Preliminary Results," in *Proceedings of the 11th NAPRECA Symposium*, 154–60 (Antananarivo, 2006).
77. Anon #3–1H.
78. CNRE, Institute de Pauster (IdP), IMRA, and partners in France and South Africa.
79. Anon #3–1I.
80. Anon #3–1J.
81. Anon #3–1K.
82. Anon #3–1L.
83. Anon #3–1M.
84. Anon #3–1N.
85. Anon #3–1O.
86. Anon #3–1P.
87. Anon #3–1Q.
88. Anon #3–1R.
89. World Bank, https://data.worldbank.org/country/MG.
90. Anon #3–1S.
91. Anon #3–1T.
92. The term "local communities" is used commonly as an area of intervention by the ICBG microprojects.
93. Anon #3–1U. The term "upfront compensation" was designed by the architects of the ICBG-Suriname project and imported into Phase I of the ICBG in Zahamena.
94. Anon #3–1V.
95. Anon #3–1W.
96. Anon #3–1X. I conducted a detailed survey of forest resource use by rural Malagasy in the Beforona area for my master's research in 1999 and 2000.

97. Anon #3–1Y.
98. Anon #3–1Z.
99. Anon #3–2A (Ariary [Ar] was calculated at the January 1, 2006, rate of exchange at roughly 2,029 for one U.S. dollar.)
100. Anon #3–2Ca.
101. Kingston, "Bioprospecting for Biodiversity Conservation."
102. Kingston, "Bioprospecting for Biodiversity Conservation."
103. Anon #3–2B.
104. Robinson, *Biodiversity, Access and Benefit-Sharing.*
105. Anon #3–2D.
106. Anon #3–2E.
107. Anon #3–2F.
108. Anon #3–2G.
109. Li, *The Will to Improve.*
110. This crude criterion relates to many parts of the world; it is not unique to Madagascar.

Chapter 4

1. Offsetting as a "license to trash" is a quote from Tom Tew, chief executive of the Environment Bank. D. Carrington, "Biodiversity Offsetting Proposals 'A Licence to Trash Nature,'" *Guardian*, September 5, 2013, https://www.theguardian.com /environment/2013/sep/05/biodiversity-offsetting-proposals-licence-to-trash. P. Greenfield, "Is a Madagascan Mine the First to Offset Its Destruction of Rainforest?" *Guardian*, March 9, 2022, https://www.theguardian.com/environment /2022/mar/09/ambatovy-the-madagascan-mine-that-might-prove-carbon-off setting-works-aoe. For a critical supplement to Julia Jones's quote, see C. Bid aud, K. Schreckenberg, M. Rabeharison, P. Ranjatson, J. Gibbons, and J. Jones, "The Sweet and the Bitter: Intertwined Positive and Negative Social Impacts of a Biodiversity Offset," *Conservation and Society* 15, no. 1 (2017): 1–13; C. Bid aud, K. Schreckenberg, and J. Jones, "The Local Costs of Biodiversity Off sets: Comparing Standards, Policy and Practice," *Land Use Policy* 77 (2018): 43–50.
2. Contracts for nickel out of Ambatovy are still for traditional appliance "white goods"; however, demand is changing to green extractives. For more on green extractives, see https://www.politico.com/news/2020/12/02/china-cobalt-mining -441967.
3. B. K. Sovacool, S. H. Ali, M. Bazilian, B. Radley, B. Nemery, J. Okatz, D. Mul vaney, "Sustainable Minerals and Metals for a Low-Carbon Future," *Science* 367, no. 6473 (2020): 30–33.
4. Informants noted that Ambatovy does not yet provide nickel and cobalt directly for green energy, but plans are certainly forthcoming, given governments' net-zero plans and the rise in demand globally for lithium-ion batteries.

5.   J. M. Klinger, *Rare Earth Frontiers: From Terrestrial Subsoils to Lunar Land-scapes* (Ithaca, N.Y.: Cornell University Press, 2018); T. Riofrancos, *Resource Radicals: From Petro-nationalism to Post-extractivism in Ecuador* (Durham: Duke University Press, 2020).

6.   There was an early report of a leaching event that resulted in four deaths and fifty hospitalizations (see https://ejatlas.org/conflict/ambatovy-mining-project -madagascar). Structural failure due to corrosion of an atmospheric leach tank's wall in the Ravenswood nickel mine in Western Australia is a case in point. In 2014, up to two megaliters (2,000 m^3) of acidic slurry exited the tank and inundated the surrounding area (see http://www.mining.com/first-quantum-dinged -40000-sulphuric-acid-spill-nickel-mine/).

7.   Ambatovy, "About Us," http://www.ambatovy.com/ambatovy-html/docs/index .html%3Fp=.html.

8.   Anon #4–1B.

9.   Such green economy programs have not been without critique. For the past three decades, the green economy has been at the forefront of global conservation and development. Green economy platforms, from direct payment to biodiversity conservation to payments ecosystem services (PES) and other market-based conservation approaches originally formalized under The Economics of Ecosystems and Biodiversity (TEEB), were delivered as part of a larger consortium of programs, such as the Global Deal for Nature and Natural Capital Accounting. See http://teebweb.org/.

10. Neimark et al. "Not Just Participation"; see also K. MacDonald, "The Devil Is in the (Bio)diversity: Private Sector 'Engagement' and the Restructuring of Biodiversity Conservation," *Antipode* 42, no. 3 (2010): 513–50.

11. Bidaud, Schreckenberg, and Jones, "The Local Costs."

12. R. Lave and M. Robertson, "Biodiversity Offsetting," in *The Routledge Handbook of the Political Economy of Science*, ed. D. Tyfield et al. (Oxon, UK: Routledge, 2017), 224–36.

13. Business and Biodiversity Offsets Programme (BBOP), *Pilot Project Case Study: The Ambatovy Project* (Washington, D.C., 2009); BBOP operates globally, but Madagascar was clearly consequential, for it also had the second pilot offsetting site in Madagascar. This site included the Rio Tinto ilmenite megamine in the south of Madagascar.

14. Based on Guy Standing's *The Precariat*—see Neimark et al., "Not Just Participation."

15. S. Sullivan, "After the Green Rush? Biodiversity Offsets, Uranium Power and the 'Calculus of Casualties' in Greening Growth," *Human Geography* 6, no. 1 (2013): 80–101, 82–83.

16. Neimark and Wilson, "Re-mining the Collections"; Seagle, "Inverting the Impacts"; Benabou, "Making Up for Lost Nature?"; E. Apostolopoulou, *Nature Swapped and Nature Lost: Biodiversity Offsetting, Urbanization and Social Justice* (Cham: Springer Nature, 2020).

17. B. Büscher, W. Dressler, and R. Fletcher, *Nature Inc.: Environmental Conservation in the Neoliberal Age* (Tucson: University of Arizona Press, 2014).

18. G. Ceballos, P. R. Ehrlich, and R. Dirzo, "Biological Annihilation via the Ongoing Sixth Mass Extinction Signalled by Vertebrate Population Losses and Declines," *Proceedings of the National Academy of Sciences* 114, no. 30 (2017): E6089–E6096; or see D. Carrington, "Earth's Sixth Mass Extinction Event Under Way, Scientists Warn," *Guardian*, July 10, 2017, https://www.theguardian .com/environment/2017/jul/10/earths-sixth-mass-extinction-event-already -underway-scientists-warn; H. Cockburn, "Earth Accelerating Towards Sixth Mass Extinction Event That Could See 'Disintegration of Civilisation,' Scientists Warn," *Independent*, June 2, 2020. https://www.independent.co.uk/environ ment/sixth-mass-extinction-endangered-animals-wildlife-markets-biodiversity -crisis-standford-study-a9544856.html.

19. E. O. Wilson, *Half-Earth: Our Planet's Fight for Life* (New York: W. W. Norton, 2016); S. Dasgupta, "Will Protecting Half the Earth Save Biodiversity? Depends Which Half," *Mongabay*, August 30, 2018, https://news.mongabay.com/2018/08 /will-protecting-half-the-earth-save-biodiversity-depends-which-half/.

20. C. Zimmer, "Bringing Them Back to Life," *National Geographic* (2013), https:// www.nationalgeographic.com/magazine/2013/04/species-revival-bringing -back-extinct-animals/.

21. Wilson, *Half-Earth*; https://www.half-earthproject.org/pledge/.

22. A. Asiyanbi and J. Lund, "Policy Persistence: REDD+ Between Stabilization and Contestation," *Journal of Political Ecology* 27, no. 1 (2020): 378–400.

23. J. Schleicher, J. G. Zaehringer, C. Fastré, B. Vira, P. Visconti, and C. Sandbrook, "Protecting Half of the Planet Could Directly Affect over One Billion People," *Nature Sustainability* 2, no. 12 (2019): 1094–96.

24. B. Büscher, R. Fletcher, D. Brockington, C. Sandbrook, W. Adams, L. Campbell, C. Corson, et al., "Half-Earth or Whole Earth? Radical Ideas for Conservation, and Their Implications," *ORYX* 51 (2017): 407–10, https://doi.org/10.1017/S00 30605316001228; H. Kopnina, "Half the Earth for People (or More)? Addressing Ethical Questions in Conservation," *Biological Conservation* 203 (2016): 176–85.

25. S. Mollett and T. Kepe, *Land Rights, Biodiversity Conservation and Justice: Rethinking Parks and People* (Oxon, UK: Routledge, 2018).

26. R. Collard and J. Dempsey, "Life for Sale? The Politics of Lively Commodities," *Environment and Planning A* 45, no. 11 (2013): 2682–99.

27. Steve Lerner expands on the concept of sacrifice zones to show that the term has geopolitical baggage: "National Sacrifice Zones" was a label used by government officials for areas contaminated by uranium mining and processing for nuclear weapons during the Cold War nuclear arms race between the United States and the Soviet Union. S. Lerner, *Sacrifice Zones: The Front Lines of Toxic Chemical Exposure in the United States* (Cambridge, Mass.: MIT Press, 2010), 3.

28. V. Kuletz, *The Tainted Desert: Environmental and Social Ruin in the American West* (New York: Routledge, 2016).

29. C. Zografos and P. Robbins, "Green Sacrifice Zones, or Why a Green New Deal Cannot Ignore the Cost Shifts of Just Transitions," *One Earth* 3, no. 5 (2020): 543–46.

30. Sarrasin, "The Mining Industry."

31. As noted above, Ambatovy's current customers are not yet sourcing nickel and cobalt for EV batteries. However, they are well aware that future market demand for their minerals will be for lithium battery use.

32. Anon #4–1C.

33. Anon #4–1D.

34. https://ambatovy.com/en/.

35. Ambatovy's estimate of its contributions to the government of Madagascar over the lifecycle of the mine, based on current nickel prices. The exact amount payable to the government is difficult to forecast due to multiple variables that include the fluctuating market price of nickel and cobalt, the cost of input commodities (such as coal, limestone, and sulfur), inflation, etc.

36. Anon #4–1E.

37. Anon #4–1F.

38. Anon #4–1G.

39. Anon #4–1H.

40. A. Brock and A. Dunlap, "Normalising Corporate Counterinsurgency: Engineering Consent, Managing Resistance and Greening Destruction Around the Hambach Coal Mine and Beyond," *Political Geography*, 62 (2018): 33–47.

41. Anon #4–1Ha.

42. S. Dickinson and P. Berner, "Ambatovy Project: Mining in a Challenging Biodiversity Setting in Madagascar," *Malagasy Nature* 3 (2010): 2–13; S. Goodman and J. Benstead, "Updated Estimates of Biotic Diversity and Endemism for Madagascar," *Oryx* 39, no. 1 (2005): 73–77.

43. Neimark and Wilson, "Re-mining the Collections," 3.

44. Anon #4–1I.

45. S. Sullivan, "At the Edinburgh Forums on Natural Capital and Natural Commons: From Disavowal to Plutonomy, via 'Natural Capital,'" December 12, 2014, http://sian sullivan.net/, 81, as quoted in Neimark and Wilson, "Re-mining the Collections," 3.

46. *Ambatovy Sustainability Report: 2020*, https://ambatovy.com/en/wp-content /uploads/2022/04/Ambatovy-Sustainability-Report-2020-EN.pdf.

47. These deposits are approximately 3 km apart and cover an area of about 1,600 ha, with depths ranging between 20 m and 100 m.

48. L. D. Ashwal and R. D. Tucker, "Geology of Madagascar: A Brief Outline," *Gondwana Research* 2, no. 3 (1999): 335–39; E. Gilli, "Volcanism-Induced Karst Landforms and Speleogenesis, in the Ankarana Plateau (Madagascar): Hypothesis and Preliminary Research," *International Journal of Speleology* 43, no. 3 (2014): 283–93, http://dx.doi.org/10.5038/1827-806X.43.3.5.

49. G. Campbell, "Gold Mining and the French Takeover of Madagascar, 1883–1914," *African Economic History* 17 (1988): 99–126.

50. A. Cooke et al., *Ambatovy Nature* (Antananarivo: Ambatovy, 2014), 6, 189.
51. Sitting alongside "megadiverse sites," such as Masoala, Montagne d'Ambre, Marojejy in the North and Ranomafana in the South. Cooke et al. *Ambatovy Nature*.
52. At this time, a mining company Dynatec (now DMC Mining), purchased the commercial rights to the mine. The ESIA was finished just prior to the sale.
53. This was initially stipulated in Article 10 of the Environment Charter of 1990. However, nowhere in the environmental charter or MECIE are biodiversity offsets stipulated. Sarrasin, "The Mining Industry."
54. Dickinson and Berner, "Ambatovy Project."
55. S. Goodman, "Biological Research Conducted in the General Andasibe Region of Madagascar with Emphasis on Enumerating the Local Biotic Diversity," *Malagasy Nature* 3 (2010): 14–34.
56. Dickinson and Berner, "Ambatovy Project."
57. Anon #4–1La.
58. Anon #4–1Lb.
59. Anon #4–1J.
60. Anon #4–1K.
61. Anon #4–1K.
62. Anon #4–1Kb.
63. Anon #4–1K.
64. Anon #4–1L.
65. Anon #4–1M.
66. At the same time as the Ambatovy program, another BBOP offsetting program of a similar scale was implemented in the south with QMM/Rio Tinto.
67. The estimated life of the operation is approximately twenty-nine years. While project construction began in 2007, it was not completed until 2012, and in January 2014 the mine finally became compliant.
68. Cooke et al., *Ambatovy Nature*.
69. N. Clark and K. Yusoff, "Geosocial Formations and the Anthropocene," *Theory, Culture and Society* 34, nos. 2–3 (2017): 3–23, 17.
70. N. Clark, personal communication, 2021.
71 Anon #4–1K.
72. Cooke et al., *Ambatovy Nature*.
73. Anon #4–N.
74. Jones et al., "Forest Conservation in Madagascar."
75. Anon #4–1O. Besides the offset areas, other areas disrupted by the mine's original construction included the tailings facility and the leaching materials plant site, which required the resettlement of two communities with over 1,200 people. The 220 km pipeline passed across a vast terrain consisting of agricultural land and rice fields. The mine estimated that the pipeline had affected 1,330 owners/farmers with 1,943 plots covering around 650 ha of rice fields. *Ambatovy Sustainability Report: 2010*, https://ambatovy.com/ang/wp-content/uploads/2021/01/Ambatovy-Sustainability-Report-2010-EN.pdf.

76. For more on the political aspects, see Li, *The Will to Improve*.
77. Leakage can also be understood as direct or indirect environmental damage because of the offset, but is mainly used to indicate environmental harm caused because populations have moved due to offset sites that are not accessible due to their new protected status.
78. Malagasy Law: Loi 96–025 in 1996; see J. Pollini, N. Hockley, and F. D. Muttenzer, "The Transfer of Natural Resource Management Rights to Local Communities," in *Conservation and Environmental Management in Madagascar*, ed. I. Scales (London: Routledge, 2014), 196–216.
79. N. Ratsifandrihamanana, "Community Management of Natural Resources: The Future of Madagascar," WWF, January 10, 2018, https://wwf.panda.org/?320350 /COURRIER2DDES2DLECTEURS2Dpar2DNanie2DRatsifandrihamanana.
80. Anon #4–1P.
81. These contractual management transfers have been described in detail in G. Andriamalala and C. J. Gardner, "Using the *Dina* Tool as Governance of Natural Resources: Lessons of Velondriake, Southwestern Madagascar," *Tropical Conservation Science* 3, no. 4 (2010): 447–64.
82. Ratsifandrihamanana, "Community Management."
83. Ratsifandrihamanana, "Community Management."
84. Jones et al., "Forest Conservation in Madagascar."
85. See Sodikoff, *Forest and Labor in Madagascar* on the social tensions around locally based surveillance of conservation areas in Madagascar.
86. Anon #4–1R.
87. Not very much different from other environmental management schemes of the world where local councils are elected to represent the population locally. However, in Madagascar these councils seem to just be adopted by conservationists and the private sector as speaking for locals, whereas they very well may not be. See Pollini, "Slash-and-Burn Cultivation," for Madagascar; see more generally K. Grove and J. Pugh, "Assemblage Thinking and Participatory Development: Potentiality, Ethics, Biopolitics," *Geography Compass* 9, no. 1(2015): 1–13.
88. Anon #4–1Sa.
89. Anon #4–1S.
90. Anon #4–1Sb.
91. Anon #4–1Sc.
92. Anon #4–1Sd.
93. Anon #4–1Se.
94. Anon #4–1Se.
95. Anon #4–1Sf.
96. Anon #4–1T.
97. Anon #4–1Sg.
98. Anon #4–1U.
99. Anon #4–1Ta.
100. Anon #4–1V.

101. Anon #4–1W.
102. Anon #4–1X.
103. Anon #4–1Y.
104. Anon #4–1Ya-1.
105. Anon #4–1Yb-2.
106. Anon #4–1Yc-3.
107. Cooke et al., *Ambatovy Nature.*

Chapter 5

1. Measured in tons of carbon dioxide equivalent (CO_2e), an offset represents the removal of one ton of carbon dioxide or its equivalent in other greenhouse gases. J. Goodward and A. Kelly, "Bottom Line on Offsets," World Resources Institute, August 1, 2010, https://www.wri.org/research/bottom-line-offsets.

2. "We cannot plant as many as 4,000 times the number of trees as existed in the pre-industrial era. There's simply not room on the planet," United Airlines CEO Scott Kirby said. To meet climate goals, the offset market will need to grow at least 15-fold by 2030, enough to absorb 23 gigatons of greenhouse gas emissions a year, according to data from the Taskforce on Scaling Voluntary Carbon Markets, a financial services group studying the issue. "The number of real, high-quality offsets across the entire market today is vanishingly small," said Jeremy Freeman, executive director of CarbonPlan, a non-profit research group. "Today's high-quality supply is not big enough for a single major corporate pledge."

3. A. Huff, "Frictitious Commodities: Virtuality, Virtue and Value in the Carbon Economy of Repair," *Environment and Planning E: Nature and Space* (2021), https://doi.org/10.1177/25148486211015056.

4. The blue economy is a platform promoting investments in the ocean and coastal marine environments, from alternative tidal and wind energy to marine tourism, deep-sea mining, transnational shipping, global waste, and most noteworthy in this instance, carbon sequestration.

5. https://www.thebluecarboninitiative.org/. See E. McLeod, G. L. Chmura, S. Bouillon, R. Salm, M. Björk, C. M. Duarte, C. Lovelock, et al., "A Blueprint for Blue Carbon: Toward an Improved Understanding of the Role of Vegetated Coastal Habitats in Sequestering CO2," *Frontiers in the Ecology and the Environment* 9, no. 10 (2011): 552–60, https://doi.org/10.1890/110004.

6. Between roughly 6–8 mg CO_2e/ha (tons of CO_2 equivalent per hectare). See also https://www.thebluecarboninitiative.org/about-blue-carbon; some estimates can be as high as ten times better annually than terrestrial forests as a carbon sink. https://oceanservice.noaa.gov/podcast/may14/mw124-bluecarbon.html.

7. M. Barbesgaard, "Blue Growth: Savior or Ocean Grabbing?" *Journal of Peasant Studies* 45, no. 1 (2018): 130–49.

8. WWF, "The Bezos Earth Fund & WWF: Investment in Community and Climate," https://www.worldwildlife.org/pages/the-bezos-earth-fund-wwf-investment-in-community-and-climate.

9. NOAA, "Coastal Blue Carbon," https://oceanservice.noaa.gov/podcast/may14/mw124-bluecarbon.html.

10. J. Childs, "Greening the Blue? Corporate Strategies for Legitimising Deep Sea Mining," *Political Geography* 74 (2019): 1–12.

11. For an excellent study of the unintended consequences of marine conservation and improved income schemes, see I. Scales, D. Friess, L. Glass, and L. Ravaoarinorotsihoarana, "Rural Livelihoods and Mangrove Degradation in South-West Madagascar: Lime Production as an Emerging Threat," *Oryx* 52, no. 4 (2017), https://doi.org/10.1017/S0030605316001630.

12. The organization has been operating in the area in cooperation with the University of Toliara's Institut Halieutique et des Sciences Marines (IHSM) and the Wildlife Conservation Society-Madagascar (WCS), who are currently focusing on developing a network of community-run Marine Protected Areas (MPAs) out of their regional headquarters in the town of Andavadoaka.

13. Andriamalala and Gardner, "Using the *Dina* Tool."

14. Offsetting was first introduced into the U.S. Clean Air Act of 1972, and then the Clean Air Act of 1990, in attempts to control the power plant pollutants NO_x and SO_2, which were causing acid rain and decimating forests in the northeastern United States.

15. Frist launched as REDD, REDD+ sought to incorporate a wider net of local and regional participants and sustainable development outcomes.

16. Trees being very efficient at sequestering atmospheric carbon through respiration.

17. REDD+ is a national voluntary program that attempts to bring "informed and meaningful involvement of all stakeholders" including "indigenous peoples and other forest-dependent communities" into its rollout and implementation. https://www.un-redd.org/.

18. D. Bumpus and D. Liverman, "Accumulation by Decarbonisation and the Governance of Carbon Offsets," *Economic Geography* 84, no. 2 (2008): 127–56.

19. J. Shankleman and A. Rathi, "Wall Street's Favorite Climate Solution Is Mired in Disagreements," *Bloomberg*, June 1, 2021, https://www.bloomberg.com/news/features/2021-06-02/carbon-offsets-new-100-billion-market-faces-disputes-over-trading-rules.

20. J. E. Vian, "Five Ways 'Green' Carbon Policies Damage Forests—and How We Can Fix the Problem," June 9, 2021, https://theconversation.com/five-ways-green-carbon-policies-damage-forests-and-how-we-can-fix-the-problem-162132?utm_source=twitter&utm_medium=bylinetwitterbutton.

21. In the history of pollution markets, two of the most well-known are the Emissions Trading Scheme (ETS) developed under the California Cap-and-Trade, and the EU Emissions Trading Scheme. Such market-based mechanisms have been around for some time, and although all the systems all differ in their structure and application, the thinking around the mechanisms is that the best way to cut emissions is to make the "polluter pay" or incentivize them to reduce their output.

22. R. Fletcher, W. Dressler, B. Büscher, and Z. R. Anderson, "Questioning REDD+ and the Future of Market-Based Conservation," *Conservation Biology* 30, no. 3 (2016): 673–75.

23. See, for example, M. M. Kansang and I. Luginaah, "Agrarian Livelihoods Under Siege: Carbon Forestry, Tenure Constraints and the Rise of Capitalist Forest Enclosures in Ghana," *World Development 113* (2019): 131–42; F. Hajdu and K. Fischer, "Problems, Causes and Solutions in the Forest Carbon Discourse: A Framework for Analysing Degradation Narratives," *Climate and Development* 9, no. 6 (2017): 537–47.

24. E. Petkova, A. Larson, and P. Pacheco, "Forest Governance, Decentralization and REDD+ in Latin America," *Forests* 1, no. 4 (2010): 250–54; J. Phelps, E. L. Webb, and A. Agrawal, "Does REDD+ Threaten to Recentralize Forest Governance?" *Science* 328, no. 5976 (2010): 312–13; C. J. Cavanagh and H. Lein, eds., "Special Section: Political Ecologies of REDD+ in Tanzania," *Journal of Eastern African Studies* 11, no. 3 (2017): 482–570.

25. A. P. Asiyanbi, "A Political Ecology of REDD+: Property Rights, Militarised Protectionism, and Carbonised Exclusion in Cross River," *Geoforum* 77 (2016): 146–56.

26. B. A. Beymer-Farris and T. J. Bassett, "The REDD Menace: Resurgent Protectionism in Tanzania's Mangrove Forests," *Global Environmental Change* 22, no. 2 (2012): 332–41.

27. A. B. Setyowati, "Governing the Ungovernable: Contesting and Reworking REDD+ in Indonesia," *Journal of Political Ecology* 27, no. 1 (2020): 456–75.

28. S. Milne, "Grounding Forest Carbon: Property Relations and Avoided Deforestation in Cambodia," *Human Ecology* 40, no. 5 (2012): 693–706, 705.

29. T. Osborne and E. Shapiro-Garza, "Embedding Carbon Markets: Complicating Commodification of Ecosystem Services in Mexico's Forests," *Annals of the American Association of Geographers* 108, no. 1 (2018): 88–105.

30. Ferguson, "REDD Comes into Fashion in Madagascar."

31. Brimont et al., "Achieving Conservation and Equity."

32. M. Poudyal, B. S. Ramamonjisoa, N. Hockley, O. S. Rakotonarivo, J. M. Gibbons, R. Mandimbiniaina, A. Rasoamanana, and J. P. Jones, "Can REDD+ Social Safeguards Reach the 'Right' People? Lessons from Madagascar," *Global Environmental Change* 37 (2016): 31–42.

33. Anon #5–1B.

34. Huff, "Frictitious Commodities."

35. H. Lovell, H. Bulkeley, and D. Liverman, "Carbon Offsetting: Sustaining Consumption?" *Environment and Planning A* 41, no. 10 (2009): 2357–79.

36. H. Lovell and D. Liverman, "Understanding Carbon Offset Technologies," *New Political Economy* 15, no. 2 (2010): 255–73.

37. Huff, "Frictitious Commodities," 4.

38. Huff, "Frictitious Commodities," 18.

39. Y. A. Collins, "Colonial Residue: REDD+, Territorialisation and the Racialized Subject in Guyana and Suriname," *Geoforum* 106 (2019): 38–47, 44.

40. W. Carton and E. Andersson, "Where Forest Carbon Meets Its Maker: Forestry-Based Offsetting as the Subsumption of Nature," *Society & Natural Resources* 30, no. 7 (2017): 829–43. Carton and Andersson (2017, 831) insist that "this process [of carbon offsetting] involves not just the appropriation or 'grabbing' of an 'already-existing' nature . . . but also the discursive and material reworking of nature so as to make it more amenable to carbon market imperatives." Carton and Andersson argue that this is accomplished through "reshaping historical landscapes, applying particular offsetting technologies, or prioritizing certain species over others."

41. D. M. Lansing, "Performing Carbon's Materiality: The Production of Carbon Offsets and the Framing of Exchange," *Environment and Planning A* 44, no. 1 (2012): 204–20.

42. H. Lovell and N. S. Ghaleigh, "Climate Change and the Professions: The Unexpected Places and Spaces of Carbon Markets," *Transactions of the Institute of British Geographers* 38, no. 3 (2013): 512–16.

43. See S. Besky and A. Blanchette, eds., *How Nature Works: Rethinking Labor on a Troubled Planet* (Albuquerque: University of New Mexico Press, 2019).

44. Ferguson, "REDD Comes into Fashion."

45. B. Büscher, "Selling Success: Constructing Value in Conservation and Development," *World Development* 57 (2014): 79–90.

46. J. Axworthy, "Africa's Blue Economy: Five Nations Poised for Growth," *Raconteur*, September 10, 2019, https://www.Raconteur.Net/Global-Business/Africa/Blue-Economy-Africa/.

47. World Bank, "What Is the Blue Economy?" June 5, 2016, https://www.worldbank.org/en/news/infographic/2017/06/06/blue-economy.

48. Blue Ventures, "Marine Conservation Championed by Madagascar President During UK Visit," November 20, 2015, https://blueventures.org/marine-conser vation-championed-by-madagascar-president-during-uk-visit-2/.

49. This was building off the previous green economy and its market-based conservation, embedded in the 2003 Durban Vision, at the IUCN World Parks Congress in Durban, where the Malagasy government pledged to triple the surface of its protected areas—i.e., increase coverage from 1.7 million ha to 6 million ha.

50. A second goal was to calculate what nations could do to seek out limits for a 1.5 degrees Celsius rise.

51. Anon #5–1C.

52. Huff, "Frictitious Commodities."

53. For this reason, it was thought of as a "boutique" or "carbon add-on."

54. Voluntary carbon markets are less structured and more flexible and use a diverse array of methodologies for certification. They are also quite a bit less transparent, and exact data is hard to come by; however, it is estimated that roughly seventy million carbon offsets were used on the voluntary carbon market in 2019 alone. This makes it much smaller in total volume than the compliance markets, as "much as 9 billion allowances on the EU-ETS alone." IUCN, *Manual for the*

Creation of Blue Carbon Projects in Europe and the Mediterranean, ed. M. Otero (N.p.: IUCN, 2021), https://www.iucn.org/resources/file/manual-creation-blue -carbon-projects-europe-and-mediterranean.

55. Some also took part in social monitoring projects, which were funded by the Global Socioeconomic Monitoring Initiative for Coastal Management (Soc-Mon), based on the former SEMP (Socioeconomic Monitoring Project). The SocMon initiative is supported by CORDIO and Sida (Swedish International Development).

56. Offset Guide, "How to Acquire Carbon Offset Credits," https://www.offset guide.org/understanding-carbon-offsets/how-to-acquire-carbon-offset-credits /#_ftnref1.

57. World Bank, "Madagascar: Balancing Conservation and Exploitation."

58. Some estimates place annual production capacity at close to $750 million, 6.6 percent of total exports and more than 7 percent of the national GDP.

59. There are two types of markets for carbon offsets, compliance and voluntary, subject to monitoring and verified certification (e.g., Verra, or formally, Verified Carbon Standard).

60. Anon #5–1Ca. The UN Environment Blue Forests Project (BFP) is an initiative of the Global Environment Facility (GEF) focused on harnessing coastal carbon (e.g., blue carbon) and associated "ecosystem services" for development and climate action.

61. The umbrella Carbon Mitigation Blue Forests Project was funded by the Darwin Initiative, through UK government funding, the Global Environment Facility (GEF), through their Blue Forests Project, and the MacArthur Foundation. The locally managed mangrove project is located within ten villages: Lamboara, Befandefa, Tampolove, Agnolignoly, Ankitambagna, Vatoavo, Ankindranoka, Ampasimara, Ankilimalinika, and Andalambezo.

62. There are seventeen marine protected areas in Madagascar, with roughly 18 percent of those comanaged by local communities.

63. Vezo are thought to be of mixed origins, derived from groups including the Sakalava in the west, Tandroy in the south, and Masikoro and Mahafaly in the southwest. See E. Fauroux, J. Laroche, and M. Marikandia, *Brève esquisse d'une description de la société Vezo (littoral occidental de Madagascar) à la fin du XXᵉ siècle* (Toliara: ERA/CNRE ORSTOM/IFSH, 1992); C. Grenier, "Genre de vie vezo, pêche 'traditionnelle' et mondialisation sur le littoral sud-ouest de Madagascar," *Annales de géographie* 693 (2013): 549–71; R. Astuti, *People of the Sea* (Cambridge: Cambridge University Press, 1995); T. A. Oliver, K. L. Oleson, H. Ratsimbazafy, D. Raberinary, S. Benbow, and A. Harris, "Positive Catch and Economic Benefits of Periodic Octopus Fishery Closures: Do Effective, Narrowly Targeted Actions 'Catalyze' Broader Management?" *PLoS One* 10, no. 6 (2015): e0129075.

64. Although generally biodiverse, two species of mangroves are very high in abundance: *Ceriops tagal* and *Rhizophora mucronata*.

65. M. Barnes-Mauthe, K. L. Oleson, and B. Zafindrasilivonona, "The Total Economic Value of Small-Scale Fisheries with a Characterization of Post-landing Trends: An Application in Madagascar with Global Relevance," *Fisheries Research* 147 (2013): 175–85.
66. Scales et al., "Rural Livelihoods and Mangrove Degradation."
67. Grenier, "Genre de vie vezo."
68. Andriamalala and Gardner, "Using the *Dina* Tool."
69. See Scales et al., "Rural Livelihoods and Mangrove Degradation."
70. T. G. Jones, L. Glass, S. Gandhi, L. Ravaoarinorotsihoarana, A. Carro, L. Benson, H. R. Ratsimba, et al. "Madagascar's Mangroves: Quantifying Nation-Wide and Ecosystem Specific Dynamics, and Detailed Contemporary Mapping of Distinct Ecosystems," *Remote Sensing* 8, no. 2 (2016): 106.
71. The credits are managed by the Plan Vivo Standard, a volunteer carbon certifier developed by the Edinburgh Centre for Carbon Management in partnership with the University of Edinburgh and other local organizations.
72. Ministère de L'Environnement, des Forêts, et des Tourisme (MEFT), *Manuel de création des Aires Marine Protégées a Madagascar* (Antannarivo: DGEF, 2009).
73. "Mangrove" may refer to all the plants in a mangrove swamp, but it is more usually used to denote the trees themselves.
74. AMNH, "Why Mangroves Matter," https://www.amnh.org/explore/videos/bio diversity/mangroves-the-roots-of-the-sea/why-mangroves-matter.
75. This is unlike terrestrial forests, where the carbon is generally found stored in the biomass of the trees themselves.
76. L. Pendleton, D. C. Donato, B. C. Murray, S. Crooks, W. A. Jenkins, S. Sifleet, C. Craft, et al. "Estimating Global 'Blue Carbon' Emissions from Conversion and Degradation of Vegetated Coastal Ecosystems," *PLoS One* 7, no. 9 (2012), https://doi.org/10.1371/journal.pone.0043542.
77. https://medium.com/environmental-science-department/a-mangrove-mis understanding-74011178d73c.
78. https://medium.com/environmental-science-department/mangroves-are-silent -sentinels-of-our-coast-that-need-our-protection-1771c1f6b9d3—Conservancy of SWFL.
79. Anon #5–1E.
80. Anon #5–1F.
81. *The Economist*, "How to Scale Up Blue-Carbon Projects," April 22, 2021, https://ocean.economist.com/blue-finance/articles/how-to-scale-up-blue-carbon-projects.
82. Anon #5–1G.
83. Anon #5–1H.
84. Anon #5–1I.
85. Anon #5–1J.
86. Anon #5–1K.
87. Blue carbon initiative webinar: "Webinar 3: Blue Forests Science for the Oceans We Want," https://us02web.zoom.us/j/82013671252.

88. Anon #5–1K.

89. Anon #5–1L.

90. Mangrove monitors are paid on daily contracts starting from 10,000Ar ($2.50), plus travel and accommodation. This is for a full day's work (8am–6pm, with a two-hour lunch). Permanent staff said they can get paid up to 100,000Ar ($19.30) per month. These salaries are quite significant, giving the lack of economic options in the region.

91. Anon #5–1La.

92. Anon #5–1M.

93. Anon #5–1N.

94. Anon #5–1O. This raises the question if this can ever truly be a "bottom-up activity." Nevertheless, the work of the patrols is important and this work is submitted as part of the project's annual reporting, and eventually into the five-year audit conducted by the certifying agency, Plan Vivo.

95. Blue Ventures, awards: https://blueventures.org/tag/awards/.

96. Anon #5–1P.

97. Anon #5–1O.

98. Anon #5–1Q.

99. Anon #5–1R.

100. Blue Ventures depends entirely on support from private donors, volunteers, research grants, and fundraising initiatives to sustain its work alongside local project partners on conservation tourism, community health, aquaculture, and blue forest carbon sequestration.

101. Anon #5–1S.

102. Anon #5–1T.

103. Anon #5–1U.

104. Anon #5–1V.

105. Anon #5–1W.

106. Anon #5–1X.

107. Anon #5–1Y.

108. Anon #5–1Z.

109. Anon #5–2A.

110. Anon #5–2B.

111. J. J. Silver, N. J. Gray, L. M. Campbell, L. W. Fairbanks, and R. L. Gruby, "Blue Economy and Competing Discourses in International Oceans Governance," *Journal of Environment and Development* 24, no. 2 (2015): 135–60; Z. Brent, M. Barbesgaard, and C. Pedersen, *The Blue Fix: Unmasking the Politics Behind the Promise of Blue Growth* (N.p.: Transnational Institute, 2018); N. J. Bennett, H. Govan, and T. Satterfield, "Ocean Grabbing," *Marine Policy* 57 (2015): 61–68, https://doi.org/10.1016/j.marpol.2015.03.026; P. Bond, "Blue Economy Threats, Contradictions and Resistances Seen from South Africa," *Journal of Political Ecology* 26, no. 1 (2019): 341–62.

112. Barbesgaard, "Blue Growth"; T. A. Benjaminsen and I. Bryceson, "Conservation, Green/Blue Grabbing and Accumulation by Dispossession in Tanzania," *Journal of Peasant Studies* 39, no. 2: 335–55, https://doi.org/10.1080/03066150.2012.667405; Bennett, Govan, and Satterfield, "Ocean Grabbing"; J. Childs and C. Hicks, eds., "Political Ecologies of the Blue Economy in Africa," *Journal of Political Ecology* 26 (2019): 323–465, https://doi.org/10.1016/j.marpol.2015.03.026.

113. Asiyanbi and Lund, "Policy Persistence."

Conclusion

1. D. Brockington, *Fortress Conservation: The Preservation of the Mkomazi Game Reserve, Tanzania* (Bloomington: Indiana University Press, 2002); D. Anderson and R. H. Grove, *Conservation in Africa: Peoples, Policies and Practice* (Cambridge: Cambridge University Press, 1989).

2. McAfee, "Selling Nature to Save It?"; Schroeder and Neumann, "Manifest Ecological Destinies"; D. Brockington, R. Duffy, and J. Igoe, *Nature Unbound: Conservation, Capitalism and the Future of Protected Areas* (London: Routledge, 2012); P. West, *Conservation Is Our Government Now: The Politics of Ecology in Papua New Guinea* (Durham, N.C.: Duke University Press, 2006); Büscher et al., "Towards a Synthesized Critique."

3. Kingston, "Successful Drug Discovery"; J. S. Miller, "Impact of the Convention on Biological Diversity: The Lessons of Ten Years of Experience with Models for Equitable Sharing of Benefits," in *Biodiversity and the Law: Intellectual Property, Biotechnology and Traditional Knowledge*, ed. C. McManis (London: Earthscan, 2007).

4. Büscher, Dressler, and Fletcher, eds., *Nature Inc.*; Corson, MacDonald, and Neimark, "Grabbing 'Green.'"

5. P. J. Ferraro and S. K. Pattanayak, "Money for Nothing? A Call for Empirical Evaluation of Biodiversity Conservation Investments," *PLoS biology* 4, no. 4 (2006): e105.

6. C. Katz, "Whose Nature, Whose Culture? Private Productions of Space and the Preservation of Nature," in *Remaking Reality: Nature at the Millennium*, ed. B. Braun and C. Castree (London: Routledge, 1998), 46–63; N. Heynen, J. McCarthy, S. Prudham, P. Robbins, eds., *Neoliberal Environments: False Promises and Unnatural Consequences* (London: Routledge, 2007); R. Fletcher, "Capitalizing on Chaos: Climate Change and Disaster Capitalism," *Ephemera: Theory and Politics in Organization* 22, no. 3 (2012).

7. So much so that, after years of poor performance, these markets remain central in most, if not all, conservation policies moving forward. For a fresh perspective on neoliberal conservation's structural tendencies to fail and be ineffective and yet still persist, see R. Fletcher, *Failing Forward: The Rise and Fall of Neoliberal Conservation* (Berkeley: University of California Press, 2023). See also Corson, MacDonald, and Neimark, "Grabbing 'Green.'"

8. Fairhead, Leach, and Scoones, "Green Grabbing."

9. S. Mahanty, *Unsettled Frontiers: Market Formation in the Cambodia-Vietnam Borderlands* (Ithaca: N.Y.: Cornell University Press, 2022); A. Huff and Y. Orengo, "Resource Warfare, Pacification and the Spectacle of 'Green' Development: Logics of Violence in Engineering Extraction in Southern Madagascar," *Political geography* 81 (2020): 102195; J. Bluwstein and J. F. Lund, "Territoriality by Conservation in the Selous–Niassa Corridor in Tanzania," *World Development* 101 (2018): 453–65.

10. J. C. Ribot, "Theorizing Access: Forest Profits along Senegal's Charcoal Commodity Chain," *Development and Change* 29, no. 2 (1998): 307–41; Fairhead, Leach, and Scoones, "Green Grabbing"; C. Zerner, "Introduction: Toward a Broader Vision of Justice and Nature Conservation," in *People, Plants, and Justice*, ed. C. Zerner (New York: Columbia University Press, 2000), 3–20; Schroeder, *Shady Practices*; Rocheleau, Thomas-Slayter, and Wangari, *Feminist Political Ecology*.

11. M. Prieto, "Indigenous Resurgence, Identity Politics, and the Anticommodification of Nature: The Chilean Water Market and the Atacameno People," *Annals of the American Association of Geographers* 112, no. 2 (2022): 487–504.

12. Rocheleau, Thomas-Slayter, and Wangari, *Feminist Political Ecology*; Schroeder, *Shady Practices*; Elmhirst, "Introducing New Feminist Political Ecologies"; Nightingale, "A Feminist in the Forest"; Sundberg, "Feminist Political Ecology."

13. N. L. Peluso and M. Watts, eds., *Violent Environments* (Ithaca, N.Y.: Cornell University Press, 2001).

14. F. Sultana, "Political Ecology 1: From Margins to Center," *Progress in Human Geography* 45, no. 1 (2021): 156–65, 156.

15. Sodikoff, *Forest and Labor in Madagascar*.

16. B. Neimark, S. Osterhoudt, H. Alter, and A. Gradinar, "A New Sustainability Model for Measuring Changes in Power and Access in Global Commodity Chains: Through a Smallholder Lens," *Palgrave Communications* 5, no. 1 (2019), https://doi.org/10.1057/s41599-018-0199-0; L. Álvarez and B. Coolsaet, "Decolonizing Environmental Justice Studies: A Latin American Perspective," *Capitalism Nature Socialism* 31, no. 2 (2020): 50–69.

17. *New York Times*, "Turning Air Into Perfume," December 12, 2021, https://www .nytimes.com/2021/12/16/fashion/air-eau-de-parfum-perfume.html; C. Jin, "Fungi Can Help Concrete Heal Its Own Cracks," *Conversation*, January 19, 2018, https://theconversation.com/fungi-can-help-concrete-heal-its-own-cracks -90375?xid=PS_smithsonian; *New York Times*, "Spider Silk Is Stronger Than Steel. It Also Assembles Itself," November 4, 2020, https://www.nytimes.com /2020/11/04/science/spider-silk-web-self-assembly.html.

18. Neimark et al., "Mob Justice and 'the Civilized Commodity.'"

19. This layering of discourse around nature as a spectacle is articulated in Igoe, *The Nature of the Spectacle*.

20. C. Lomas, "Mining the Moon: Earth's Back-Up Plan?" *Deutsche Welle*, February 2, 2018, https://www.dw.com/en/mining-the-moon-earths-back-up-plan/a-42286299.

21. Similar to the anecdote described in chapter 3, when excitement surrounding an untested Malagasy tincture as a cure for COVID-19 and the drive to find a miracle cure was on everybody's minds.

22. See https://wildlifetradefutures.com/.

23. https://www.theguardian.com/environment/2021/sep/29/covid-tests-and-superbugs-how-the-deep-sea-could-help-us-fight-pandemics. This is spurred by the discovery of Griffithsin *Griffithsia corallina*, a lectin found from red algae. See C. Lee, "Griffithsin, a Highly Potent Broad-Spectrum Antiviral Lectin from Red Algae: From Discovery to Clinical Application," *Marine Drugs* 17, no. 10 (2019): 567.

24. N. Merino, H. S. Aronson, D. P. Bojanova, J. Feyhl-Buska, M. L. Wong, S. Zhang, and D. Giovannelli, "Living at the Extremes: Extremophiles and the Limits of Life in a Planetary Context," *Frontiers in Microbiology* 10 (2019): 780.

25. https://www.theatlantic.com/magazine/archive/2020/01/20000-feet-under-the-sea/603040/.

26. Sullivan, "Banking Nature?"

27. D. Banoub, G. Bridge, B. Bustos, I. Ertör, M. González-Hidalgo, and J. A. de los Reyes, "Industrial Dynamics on the Commodity Frontier: Managing Time, Space and Form in Mining, Tree Plantations and Intensive Aquaculture," *Environment and Planning E: Nature and Space* 4, no. 4 (2021): 1533–59.

28. Fletcher, *Failing Forward*.

29. A. L. Tsing, *Friction: An Ethnography of Global Connection* (Princeton, N.J.: Princeton University Press, 2011). See also Mahanty, *Unsettled Frontiers*.

30. Le Billon, "Crisis Conservation and Green Extraction," 865.

31. J. Dempsey and R. C. Collard, "If Biodiversity Offsets Are a Dead End for Conservation, What Is the Live Wire? A Response to Apostolopoulou & Adams," *Oryx* 51, no. 1 (2017): 35–39, http://doi.org/10.1017/S0030605316000752; J. Dempsey and D. C. Suarez, "Arrested Development? The Promises and Paradoxes of 'Selling Nature to Save It,'" *Nature and Society* 106, no. 3 (2016): 653–71; B. Büscher and V. Davidov, "Environmentally Induced Displacements in the Ecotourism–Extraction Nexus," *Area* 48, no. 2 (2016): 161–67; C. B. Enns, A. Bersaglio, and A. Sneyd, "Fixing Extraction Through Conservation: On Crises, Fixes and the Production of Shared Value and Threat," *Environment and Planning E: Nature and Space* 2, no. 4 (2019): 967–88; Huff and Orengo, "Resource Warfare."

32. Büscher and Davidov, "Environmentally Induced Displacements." For the blue economy, see I. Ertör and M. Hadjimichael, "Blue Degrowth and the Politics of the Sea: Rethinking the Blue Economy," *Sustainability Science* 15, no. 1 (2020): 1–10; Seagle, "Inverting the Impacts."

33. Neimark and Wilson, "Re-mining the Collections."

34. D. MacKenzie, "Finding the Ratchet: The Political Economy of Carbon Trading," *Post Autistic Economics Review* 42 (2007): 8–17.

35. L. Campling and A. Colás, "Capitalism and the Sea: Sovereignty, Territory and Appropriation in the Global Ocean," *Environment and Planning D: Society and Space* 36, no. 4 (2018): 776–94.

36. C. Jacob, J. W. van Bochove, S. Livingstone, T. White, J. Pilgrim, and L. Bennun, "Marine Biodiversity Offsets: Pragmatic Approaches Toward Better Conservation Outcomes," *Conservation Letters* 13, no. 3 (2020): e12711.

37. Sometimes thought of as climate change or adaptation labor.

38. See Neimark and Vermeylen, "A Human Right to Science?"; S. Mahanty and B. Neimark, "The Green Gig Economy: Precarious Workers Are on the Frontline of Climate Change Fight," *Conversation*, April 21, 2020, https://theconversation.com/the-green-gig-economy-precarious-workers-are-on-the-frontline-of-climate-change-fight-133392.

39. See Robinson, *Biodiversity, Access and Benefit-Sharing.*

40. K. Devenish, S. Desbureaux, S. Willcock, and J. P. Jones, "On Track to Achieve No Net Loss of Forest at Madagascar's Biggest Mine," *Nature Sustainability* 5 (2022): 498–508.

41. Bidaud et al., "The Sweet and the Bitter."

42. Jones, Rakotonarivo, and Razafimanahaka, "Forest Conservation in Madagascar."

43. Devenish et al., "On Track."

44. F. Müller, T. Johanna, and T. Kalt, "Hydrogen Justice," *Environmental Research Letters* 17, no. 11 (2022): 115006; C. Inverardi-Ferri, "The Enclosure of 'Waste Land': Rethinking Informality and Dispossession," *Transactions of the Institute of British Geographers* 43, no. 2 (2018): 230–44. On climate adaptation, see Mahanty and Neimark, "Green Gig Economy." See also M. Mikulewicz, L. Johnson, M. Mills-Novoa and B. Neimark, "Understanding Adaptation as Labor: Implications for Critical Climate Research." Paper presentation at the American Association of Geographers (AAG) Annual Meeting, Denver, US, March 25, 2023, https://aag.secureplatform.com/aag2023/solicitations/39/sessiongallery/6604.

45. Following recent calls for radical alternatives by critical scholars: A. Dunlap and S. Sullivan, "A Faultline in Neoliberal Environmental Governance Scholarship? Or, Why Accumulation-by-Alienation Matters," *Environment and Planning E: Nature and Space* 3, no. 2 (2020): 552–79.

46. F. Sultana, "Decolonizing Development Education and the Pursuit of Social Justice," *Human Geography* 12, no. 3 (2019): 31–46, 33.

47. W. D. Mignolo, "Delinking: The Rhetoric of Modernity, the Logic of Coloniality and the Grammar of De-coloniality," *Cultural Studies* 21, nos. 2–3 (2007): 449–514.

48. For a reading list that can inform on and assist the approach, see S. E. Cannon, "Decolonizing Conservation: A Reading List," *Zenodo*, https://doi.org/10.5281/zenodo.4429220.

49. https://conviva-research.com/what-do-we-mean-by-decolonizing-conservation -a-response-to-lanjouw-2021/.

50. See keynote lecture by P. West, "Critical Approaches to Dispossession in the Melanesian Pacific: Conservation, Voice, and Collaboration." Presentation delivered at the Second Biennial Conference of the Political Ecology Network (POLLEN), Political Ecology, the Green Economy, and Alternative Sustainabilities, Oslo Metropolitan University, Oslo, Norway, June, 19–22, 2018, https:// politicalecologynetwork.org/2018/09/01/paige-west-keynote-lecture-at-pollen 18-conference-oslo-20-june-2018/.

51. Sultana, "Political Ecology 1"; E. Baglioni, "Straddling Contract and Estate Farming: Accumulation Strategies of Senegalese Horticultural Exporters," *Journal of Agrarian Change* 15, no. 1 (2015): 17–42.

52. W. Harcourt and I. L. Nelson, eds., *Practising Feminist Political Ecologies: Moving Beyond the "Green Economy"* (London: Bloomsbury, 2015), 7; D. Haraway, "Situated Knowledges: The Science Question in Feminism and the Privilege of Partial Perspective," *Feminist Studies* 14, no. 3 (1988): 575–99.

53. Sultana, "Political Ecology 1."

54. K. Yusoff, *A Billion Black Anthropocenes or None* (Minneapolis: University of Minnesota Press, 2018).

55. P. Robbins, *Political Ecology: A Critical Introduction*, 3rd ed. (Hoboken, N.J.: John Wiley & Sons, 2019).

56. Álvarez and Coolsaet, "Decolonizing Environmental Justice Studies"; A. Martin, B. Coolsaet, E. Corbera, N. M. Dawson, J. A. Fraser, J. A., I. Lehmann, and I. Rodriguez, "Justice and Conservation: The Need to Incorporate Recognition," *Biological Conservation* 197 (2016): 254–61; J. A. Fraser, "Amazonian Struggles for Recognition," *Transactions of the Institute of British Geographers* 43, no. 4 (2018): 718–32.

57. B. Büscher and R. Fletcher, "Towards Convivial Conservation," *Conservation and Society* 17, no. 3 (2019): 283–96.

58. Büscher and Fletcher, "Towards Convivial Conservation," 289.

59. G. D'Alisa, F. Demaria, and G. Kallis, *Degrowth: A Vocabulary for a New Era* (Abingdon, Oxon, UK: Routledge, 2014).

60. https://www.tandfonline.com/doi/full/10.1080/14747731.2020.1812222.

61. G. Kallis, "In Defence of Degrowth," *Ecological Economics* 70, no. 5 (2011): 873–880, 873.

62. Kallis, "In Defence of Degrowth."

63. P. Sibanda, "The Dimensions of 'Hunhu/Ubuntu'(Humanism in the African sense): The Zimbabwean Conception," *Dimensions* 4, no. 1 (2014): 26–28; C. B. Gade, *A Discourse on African Philosophy: A New Perspective on Ubuntu and Transitional Justice in South Africa* (Lanham, Md.: Lexington Books, 2017).

64. S. Caria and R. Domínguez, "Ecuador's Buen Vivir: A New Ideology for Development," *Latin American Perspectives* 43, no. 1 (2016): 18–33; E. Gudynas, "Buen Vivir: Today's Tomorrow," *Development* 54, no. 4 (2011): 441–47.

65. Anon #6–1A.
66. Anon #6–1B.
67. Anon #6–1C.
68. Anon #6–1D.
69. Anon #6–1E.
70. Anon #6–1F.

Bibliography

Adams, C. "The DJ and the Miracle Cure: The Perils of Postcolonial Medicine in Madagascar." *Drift*, October 21, 2020, https://www.thedriftmag.com/the-dj-and-the-miracle-cure/.

Allen, P. M., and M. Covell. *Historical Dictionary of Madagascar*, Historical Dictionaries of Africa, no. 98. Lanham, Md.: Scarecrow Press, 2005.

Alpers, E. A. "Recollecting Africa: Diasporic Memory in the Indian Ocean World." *African Studies Review* 43, no. 1 (2000): 83–99.

Althabe, G. *Oppression et liberation dans l'imaginaire, les communautés villageoises le la côte orientale de Madagascar*. Paris: F. Maspero, 1969.

Álvarez, L., and B. Coolsaet. "Decolonizing Environmental Justice Studies: A Latin American Perspective." *Capitalism Nature Socialism* 31, no. 2 (2020): 50–69.

Ambatovy. "About Us," http://www.ambatovy.com/ambatovy-html/docs/index.html %3Fp=.html.

Ambatovy Sustainability Report: 2020, https://ambatovy.com/en/wp-content/uploads/2022/04/Ambatovy-Sustainability-Report-2020-EN.pdf.

AMNH. "Why Mangroves Matter," https://www.amnh.org/explore/videos/biodiversity/mangroves-the-roots-of-the-sea/why-mangroves-matter.

Anderson, D., and R. H. Grove. *Conservation in Africa: Peoples, Policies and Practice*. Cambridge: Cambridge University Press, 1989.

Anderson, T. "Solving Madagascar: Science, Illustrations, and the Normalizing of Fauna of Nineteenth Century Madagascar." In *Contest for Land in Madagascar: Environment, Ancestors and Development*, ed. S. Evers, G. Campbell, and M. Lambek, 97–118. Leiden: Brill, 2013.

Andriamalala, G., and C. J. Gardner, "Using the *Dina* Tool as Governance of Natural Resources: Lessons of Velondriake, Southwestern Madagascar," *Tropical Conservation Science* 3, no. 4 (2010): 447–64.

Andriamialisoa, F., and O. Langrand. "The History of Zoological Exploration of Madagascar." In *The Natural History of Madagascar*, ed. S. Goodman and J. Benstead,
1–15. Chicago: University of Chicago Press, 2003.

Andrianirina-Ratsialonana, R., L. Ramarojohn, P. Burnod, and A. Teyssier. *After Daewoo? Current Status and Perspectives of Large-Scale Land Acquisition in Madagascar* (2011), https://agritrop.cirad.fr/560265/1/document_560265.pdf.

Apostolopoulou, E. *Nature Swapped and Nature Lost: Biodiversity Offsetting, Urbanization and Social Justice.* Cham: Springer Nature, 2020.

Arsel, M., and B. Büscher. "Nature™ Inc.: Changes and Continuities in Neoliberal
Conservation and Market-Based Environmental Policy." *Development and Change*
43, no. 1 (2012): 53–78.

Ashwal, L. D., and R. D. Tucker. "Geology of Madagascar: A Brief Outline." *Gondwana Research* 2, no. 3 (1999): 335–39.

Asiyanbi, A., and J. Lund. "Policy Persistence: REDD+ Between Stabilization and
Contestation." *Journal of Political Ecology* 27, no. 1 (2020): 378–400.

Asiyanbi, A. P. "A Political Ecology of REDD+: Property Rights, Militarised Protectionism, and Carbonised Exclusion in Cross River." *Geoforum* 77 (2016): 146–56.

Astuti, R. *People of the Sea.* Cambridge: Cambridge University Press, 1995.

Axworthy, J. "Africa's Blue Economy: Five Nations Poised for Growth." *Raconteur*,
September 10, 2019, https://www.Raconteur.Net/Global-Business/Africa/Blue
-Economy-Africa/.

Baglioni, E. "Straddling Contract and Estate Farming: Accumulation Strategies of Senegalese Horticultural Exporters." *Journal of Agrarian Change* 15, no. 1 (2015): 17–42.

Bakker, K. J. *An Uncooperative Commodity: Privatizing Water in England and Wales.*
Oxford: Oxford Geographical and Environmental Studies, 2003.

Balick, M. J. "Ethnobotany and the Identification of Therapeutic Agents from the
Rainforest." In *Ciba Foundation Symposium 154: Bioactive Compounds from Plants*,
22–39. Chichester, UK: John Wiley & Sons, 2007.

Banoub, D., G. Bridge, B. Bustos, I Ertör, M. González-Hidalgo, and J. A. de los Reyes.
"Industrial Dynamics on the Commodity Frontier: Managing Time, Space and
Form in Mining, Tree Plantations and Intensive Aquaculture." *Environment and
Planning E: Nature and Space* 4, no. 4 (2021): 1533–59.

Barbesgaard, M. "Blue Growth: Savior or Ocean Grabbing?" *Journal of Peasant Studies* 45, no. 1 (2018): 130–49.

Barnes-Mauthe, M., K. L. Oleson, and B. Zafindrasilivonona. "The Total Economic
Value of Small-Scale Fisheries with a Characterization of Post-landing Trends:
An Application in Madagascar with Global Relevance." *Fisheries Research* 147
(2013): 175–85.

BBC. "Coronavirus: Caution Urged over Madagascar's 'Herbal Cure,'" April 22, 2020,
https://www.bbc.co.uk/news/world-africa-52374250.

Benabou, S. "Making Up for Lost Nature? A Critical Review of the International
Development of Voluntary Biodiversity Offsets." *Environment and Society* 5, no. 1
(2014): 103–23.

Benjaminsen, T. A., and I. Bryceson. "Conservation, Green/Blue Grabbing and Accumulation by Dispossession in Tanzania." *Journal of Peasant Studies* 39, no. 2: 335–55, https://doi.org/10.1080/03066150.2012.667405.

Bennett, N. J., H. Govan, and T. Satterfield. "Ocean Grabbing." *Marine Policy* 57 (2015): 61–68, https://doi.org/10.1016/j.marpol.2015.03.026.

Bertrand, A., and M. Sourdat. *Feux et Déforestation á Madagascar: Revues Bibliographiques*. Antananarivo: CIRAD/ORSTOM/CITE, 1998.

Besky, S., and A. Blanchette, eds. *How Nature Works: Rethinking Labor on a Troubled Planet*. Albuquerque: University of New Mexico Press, 2019.

Beymer-Farris, B. A., and T. J. Bassett. "The REDD Menace: Resurgent Protectionism in Tanzania's Mangrove Forests." *Global Environmental Change* 22, no. 2 (2012): 332–41.

Bidaud, C., K. Schreckenberg, and J. Jones. "The Local Costs of Biodiversity Offsets: Comparing Standards, Policy and Practice." *Land Use Policy* 77 (2018): 43–50.

Bidaud, C., K. Schreckenberg, M. Rabeharison, P. Ranjatson, J. Gibbons, and J. Jones. "The Sweet and the Bitter: Intertwined Positive and Negative Social Impacts of a Biodiversity Offset." *Conservation and Society* 15, no. 1 (2017): 1–13.

Bigger, P., J. Dempsey, A. P. Asiyanbi, K. Kay, R. Lave, B. Mansfield, T. Osborne, et al. "Reflecting on Neoliberal Natures: An Exchange." *Environment and Planning E: Nature and Space* 1, nos. 1–2 (2018): 25–75.

Bigger, P., and M. Robertson. "Value Is Simple. Valuation Is Complex." *Capitalism Nature Socialism* 28, no. 1 (2017): 68–77.

Birch, K., and D. Tyfield. "Theorizing the Bioeconomy: Biovalue, Biocapital, Bioeconomics or . . . What?" *Science, Technology & Human Values* 38, no. 3 (2013): 299–327.

Birch, K., L. Levidow, and T. Papaioannou. "Sustainable Capital? The Neoliberalization of Nature and Knowledge in the European 'Knowledge-Based Bio-economy.'" *Sustainability* 2, no. 9 (2010): 2898–918, 2899.

Blue Ventures. "Marine Conservation Championed by Madagascar President During UK Visit," November 20, 2015, https://blueventures.org/marine-conservation-championed-by-madagascar-president-during-uk-visit-2/.

Bluwstein, J., and J. F. Lund. "Territoriality by Conservation in the Selous–Niassa Corridor in Tanzania." *World Development* 101 (2018): 453–65.

Bond, P. "Blue Economy Threats, Contradictions and Resistances Seen from South Africa." *Journal of Political Ecology* 26, no. 1 (2019): 341–62.

Bracking, S. "Financialization and the Environmental Frontier." In *The Routledge International Handbook of Financialization*, ed. D. Mertens, N. van der Zwan, and P. Mader, 213–23. Abingdon, Oxon, UK: Routledge, 2020.

Brent, Z., M. Barbesgaard, and C. Pedersen. *The Blue Fix: Unmasking the Politics Behind the Promise of Blue Growth*. N.p.: Transnational Institute, 2018.

Brimont, L., D. Ezzine-de-Blas, A. Karsenty, and A. Toulon. "Achieving Conservation and Equity Amidst Extreme Poverty and Climate Risk: The Makira REDD+ Project in Madagascar." *Forests* 6, no. 3 (2015): 748–68.

Brock, A., and A. Dunlap. "Normalising Corporate Counterinsurgency: Engineering Consent, Managing Resistance and Greening Destruction Around the Hambach Coal Mine and Beyond." *Political Geography*, 62 (2018): 33–47.

Brockington, D. *Fortress Conservation: The Preservation of the Mkomazi Game Reserve, Tanzania.* Bloomington: Indiana University Press, 2002.

Brockington, D., and R. Duffy, eds. *Capitalism and Conservation.* Malden, Mass.: John Wiley & Sons, 2011.

Brockington, D., R. Duffy, and J. Igoe. *Nature Unbound: Conservation, Capitalism and the Future of Protected Areas.* London: Routledge, 2012.

Brockway, L. "Science and Colonial Expansion: The Role of the British Royal Botanic Gardens." *American Ethnologist* 6, no. 3 (1979): 449–65.

Brown, M. "Madagascar: Island of the Ancestors." *Anthropology Today* 3, no. 1 (1987): 14–17.

Bullard, N. "BlackRock's New Morality Marks the End for Coal." *Bloomberg*, January 17, 2020, https://www.bloomberg.com/opinion/articles/2020-01-17/blackrock-s-climate-conscious-tidal-wave-breaks-on-coal.

Bumpus, D., and D. Liverman. "Accumulation by Decarbonisation and the Governance of Carbon Offsets." *Economic Geography* 84, no. 2 (2008): 127–56.

Burnod, P., M. Gingembre, and R. Andrianirina Ratsialonana. "Competition Over Authority and Access: International Land Deals in Madagascar." *Development and Change* 44, no. 2 (2013): 357–79.

Büscher, B. "Letters of Gold: Enabling Primitive Accumulation Through Neoliberal Conservation." *Human Geography* 2, no. 3 (2009): 91–94.

Büscher, B. "Payments for Ecosystem Services as Neoliberal Conservation: (Reinterpreting) Evidence from the Maloti-Drakensberg, South Africa." *Conservation and Society* 10, no. 1 (2012): 29–41.

Büscher, B. "Selling Success: Constructing Value in Conservation and Development." *World Development* 57 (2014): 79–90.

Büscher, B., and V. Davidov. "Environmentally Induced Displacements in the Ecotourism–Extraction Nexus." *Area* 48, no. 2 (2016): 161–67.

Büscher, B., W. Dressler, and R. Fletcher, eds. *Nature Inc.: Environmental Conservation in the Neoliberal Age.* Tucson: University of Arizona Press, 2014.

Büscher, B., and R. Fletcher. "Accumulation by Conservation." *New Political Economy* 20, no. 2 (2015): 273–98.

Büscher, B., and R. Fletcher. "Towards Convivial Conservation." *Conservation and Society* 17, no. 3 (2019): 283–96.

Büscher, B., R. Fletcher, D. Brockington, C. Sandbrook, W. Adams, L. Campbell, C. Corson, et al. "Half-Earth or Whole Earth? Radical Ideas for Conservation, and Their Implications." *ORYX* 51 (2017): 407–10, https://doi.org/10.1017/S0030605316001228.

Büscher, B., S. Sullivan, K. Neves, J. Igoe, and D. Brockington. "Towards a Synthesized Critique of Neoliberal Biodiversity Conservation." *Capitalism Nature Socialism* 23, no. 2 (2012): 4–30.

Business and Biodiversity Offsets Programme (BBOP). *Pilot Project Case Study: The Ambatovy Project*. Washington, D.C., 2009.

Callet, R. *Tantaran'ny Andriana eto Madagascar*. Antananarivo: Académie Malgache, 1974.

Campbell, G. *An Economic History of Imperial Madagascar, 1750–1895: The Rise and Fall of an Island Empire*. Cambridge: Cambridge University Press, 2005.

Campbell, G. "Forest Depletion in Imperial Madagascar, c. 1790–1861." In *Contest for Land in Madagascar: Environment, Ancestors and Development*, ed. S. Evers, G. Campbell, and M. Lambek, 63–95. Leiden: Brill, 2013.

Campbell, G. "Gold Mining and the French Takeover of Madagascar, 1883–1914." *African Economic History* 17 (1988): 99–126.

Campbell, G. "Madagascar and Mozambique in the Slave Trade of the Western Indian Ocean, 1800–1861." *Slavery and Abolition* 9, no. 3 (1988): 165–92.

Campbell, G. "The Structure of Trade in Madagascar, 1750–1810." *International Journal of African Historical Studies* 26, no. 1 (1993): 111–48.

Campbell, G., ed. *Bondage and the Environment in the Indian Ocean World*. Cham: Springer, 2018.

Campling, L., and A. Colás. "Capitalism and the Sea: Sovereignty, Territory and Appropriation in the Global Ocean." *Environment and Planning D: Society and Space* 36, no. 4 (2018): 776–94.

Cannon, S. E. "Decolonizing Conservation: A Reading List." *Zenodo*, https://doi.org /10.5281/zenodo.4429220.

Cao, S., and D. G. Kingston. "Biodiversity Conservation and Drug Discovery: Can They Be Combined? The Suriname and Madagascar Experiences." *Pharmaceutical Biology* 47, no. 8 (2009): 809–23.

Caria, S., and R. Domínguez. "Ecuador's Buen Vivir: A New Ideology for Development." *Latin American Perspectives* 43, no. 1 (2016): 18–33.

Carrington, D. "Biodiversity Offsetting Proposals 'A Licence to Trash Nature.'" *Guardian*, September 5, 2013, https://www.theguardian.com/environment/2013 /sep/05/biodiversity-offsetting-proposals-licence-to-trash.

Carrington, D. "Earth's Sixth Mass Extinction Event Under Way, Scientists Warn." *Guardian*, July 10, 2017, https://www.theguardian.com/environment/2017/jul/10 /earths-sixth-mass-extinction-event-already-underway-scientists-warn.

Carton, W., and E. Andersson. "Where Forest Carbon Meets Its Maker: Forestry-Based Offsetting as the Subsumption of Nature." *Society & Natural Resources* 30, no. 7 (2017): 829–43.

Cavanagh, C. J., and H. Lein, eds. "Special Section: Political Ecologies of REDD + in Tanzania." *Journal of Eastern African Studies* 11, no. 3 (2017): 482–570.

Ceballos, G., P. R. Ehrlich, and R. Dirzo. "Biological Annihilation via the Ongoing Sixth Mass Extinction Signalled by Vertebrate Population Losses and Declines." *Proceedings of the National Academy of Sciences* 114, no. 30 (2017): E6089–E6096.

Chernock, A. "Queen Victoria and the 'Bloody Mary of Madagascar.'" *Victorian Studies* 55, no. 3 (2013): 425–49.

Childs, J. "Greening the Blue? Corporate Strategies for Legitimising Deep Sea Mining." *Political Geography* 74 (2019): 1–12.

Childs, J., and C. Hicks, eds. "Political Ecologies of the Blue Economy in Africa." *Journal of Political Ecology* 26 (2019): 323–465, https://doi.org/10.1016/j.marpol .2015.03.026.

Clark, N., and K. Yusoff. "Geosocial Formations and the Anthropocene." *Theory, Culture and Society* 34, nos. 2–3 (2017): 3–23.

Cockburn, H. "Earth Accelerating Towards Sixth Mass Extinction Event That Could See 'Disintegration of Civilisation,' Scientists Warn." *Independent,* June 2, 2020. https://www.independent.co.uk/environment/sixth-mass-extinction-endan gered-animals-wildlife-markets-biodiversity-crisis-standford-study-a9544856 .html.

Cole, J. *Forget Colonialism? Sacrifice and the Art of Memory in Madagascar.* Berkeley: University of California Press, 2001.

Collard, R., and J. Dempsey. "Life for Sale? The Politics of Lively Commodities." *Environment and Planning A* 45, no. 11 (2013): 2682–99.

Collins, Y. A. "Colonial Residue: REDD+, Territorialisation and the Racialized Subject in Guyana and Suriname." *Geoforum* 106 (2019): 38–47.

The Conversation. "Three Financial Firms Could Change the Direction of the Climate Crisis—and Few People Have Any Idea," February 24, 2020, https://theconver sation.com/three-financial-firms-could-change-the-direction-of-the-climate -crisis-and-few-people-have-any-idea-131869.

Cooke, A., et al. *Ambatovy Nature.* Antananarivo: Ambatovy, 2014.

Corson, C. A. *Corridors of Power: The Politics of Environmental Aid to Madagascar.* New Haven: Yale University Press, 2016.

Corson, C., K. I. MacDonald, and B. Neimark. "Grabbing 'Green': Markets, Environmental Governance and the Materialization of Natural Capital." *Human Geography* 6, no. 1 (2013): 1–15.

Cragg, G., and D. Newman. "Natural Product Drug Discovery in the Next Millennium." *Pharmaceutical Biology* 39 (2001): 8–17.

Cragg, G. M., and D. J. Newman. "Biodiversity: A Continuing Source of Novel Drug Leads." *Pure and Applied Chemistry* 77, no. 1 (2005): 7–24.

D'Alisa, G., F. Demaria, and G. Kallis. *Degrowth: A Vocabulary for a New Era.* Abingdon, Oxon, UK: Routledge, 2014.

Dasgupta, S. "Will Protecting Half the Earth Save Biodiversity? Depends Which Half." *Mongabay,* August 30, 2018, https://news.mongabay.com/2018/08/will-protec ting-half-the-earth-save-biodiversity-depends-which-half/.

Dempsey, J. *Enterprising Nature: Economics, Markets, and Finance in Global Biodiversity Politics.* West Sussex, UK: John Wiley & Sons, 2016.

Dempsey, J., and R. C. Collard. "If Biodiversity Offsets Are a Dead End for Conservation, What Is the Live Wire? A Response to Apostolopoulou & Adams." *Oryx* 51, no. 1 (2017): 35–39, http://doi.org/10.1017/S0030605316000752.

Dempsey, J., and D. C. Suarez. "Arrested Development? The Promises and Paradoxes of 'Selling Nature to Save It.'" *Nature and Society* 106, no. 3 (2016): 653–71.

Devenish, K., S. Desbureaux, S. Willcock, and J. P. Jones. "On Track to Achieve No Net Loss of Forest at Madagascar's Biggest Mine." *Nature Sustainability* 5 (2022): 498–508.

Dewar, R. E., and A. F. Richard. "Madagascar: A History of Arrivals, What Happened, and Will Happen Next." *Annual Review of Anthropology* 41 (2012): 495–517.

Dickinson, S., and P. Berner. "Ambatovy Project: Mining in a Challenging Biodiversity Setting in Madagascar." *Malagasy Nature* 3 (2010): 2–13.

Drews, J. "Drug Discovery: A Historical Perspective." *Science* 287, no. 5460 (2000): 1960–64.

Duffy, R. "Global Environmental Governance and the Politics of Ecotourism in Madagascar." *Journal of Ecotourism* 5, nos. 1–2 (2006): 128–44.

Duffy, R. "Non-governmental Organisations and Governance States: The Impact of Transnational Environmental Management Networks in Madagascar." *Environmental Politics* 15, no. 5 (2006): 731–49.

Dunlap, A., and S. Sullivan. "A Faultline in Neoliberal Environmental Governance Scholarship? Or, Why Accumulation-by-Alienation Matters." *Environment and Planning E: Nature and Space* 3, no. 2 (2020): 552–79.

The Economist. "How to Scale Up Blue-Carbon Projects," April 22, 2021, https://ocean .economist.com/blue-finance/articles/how-to-scale-up-blue-carbon-projects.

Eisner, T. "Chemical Prospecting: A Global Imperative." *Proceedings of the American Philosophical Society* 138, no. 3 (1994): 385–93.

Elmhirst, R. "Introducing New Feminist Political Ecologies." *Geoforum* 42, no. 2 (2011): 129–32.

Enns, C. B., A. Bersaglio, and A. Sneyd. "Fixing Extraction Through Conservation: On Crises, Fixes and the Production of Shared Value and Threat." *Environment and Planning E: Nature and Space* 2, no. 4 (2019): 967–88.

Ertör, I., and M. Hadjimichael. "Blue Degrowth and the Politics of the Sea: Rethinking the Blue Economy." *Sustainability Science* 15, no. 1 (2020): 1–10.

European Commission. "New Perspectives on the Knowledge-Based Bio-Economy." Conference Report. Brussels: DG-Research, 2005.

European Commission. "Report on Innovating for Sustainable Growth: A Bioeconomy for Europe," #A7–0201/2013, Brussels, 2013, https://www.europarl.europa .eu/doceo/document/A-7-2013-0201_EN.html.

Evers, S., G. Campbell, and M. Lambek. "Land Competition and Human-Environment Relations in Madagascar." In *Contest for Land in Madagascar: Environment, Ancestors and Development*, ed. S. Evers, G. Campbell, and M. Lambek, 1–20. Leiden: Brill, 2013.

Fairhead, J., and M. Leach. *Misreading the African Landscape: Society and Ecology in a Forest-Savanna Mosaic.* Cambridge: Cambridge University Press, 1996.

Fairhead, J., M. Leach, and I. Scoones. "Green Grabbing: A New Appropriation of Nature?" *Journal of Peasant Studies* 39, no. 2 (2012): 237–61.

Fauroux, E., J. Laroche, and M. Marikandia. *Brève esquisse d'une description de la société Vezo (littoral occidental de Madagascar) à la fin du XXe siècle.* Toliara: ERA/CNRE ORSTOM/IFSH, 1992.

Feeley-Harnik, G. *Green Estate Restoring Independence in Madagascar*. Washington, D.C.: Smithsonian, 1991.

Ferguson, B. "REDD Comes into Fashion in Madagascar." *Madagascar Conservation & Development* 4, no. 2 (2009), https://www.ajol.info/index.php/mcd/article/view /48654.

Ferraro, P. J., and S. K. Pattanayak. "Money for Nothing? A Call for Empirical Evaluation of Biodiversity Conservation Investments." *PLoS biology* 4, no. 4 (2006): e105.

Fletcher, R. "Capitalizing on Chaos: Climate Change and Disaster Capitalism." *Ephemera: Theory and Politics in Organization* 22, no. 3 (2012).

Fletcher, R. *Failing Forward: The Rise and Fall of Neoliberal Conservation*. Berkeley: University of California Press, 2023.

Fletcher, R., W. Dressler, B. Büscher, and Z. R. Anderson. "Questioning REDD+ and the Future of Market-Based Conservation." *Conservation Biology* 30, no. 3 (2016): 673–75.

"Forest Carbon Partnerships, Readiness Package (R-Package) for Reducing Emissions from Deforestation and Forest Degradation in Madagascar," 2017, https://www .forestcarbonpartnership.org/system/files/documents/Madagascar-TAP%20R -Package%20Review-10%20Septj.pdf.

Forsyth, T. *Critical Political Ecology: The Politics of Environmental Science*. Abingdon, Oxon, UK: Routledge, 2004.

Fraser, J. A. "Amazonian Struggles for Recognition." *Transactions of the Institute of British Geographers* 43, no. 4 (2018): 718–32.

Freudenberger, K. S., and the International Resources Group. *Paradise Lost? Lesson from 25 Years of USAID Environment Programs in Madagascar*. Washington, D.C.: U.S. Agency for International Development, 2010.

Gade, C. B. *A Discourse on African Philosophy: A New Perspective on Ubuntu and Transitional Justice in South Africa*. Lanham, Md.: Lexington Books, 2017.

Ganzhorn, J., U. Porter, P. Lowry, G. E. Schatz, and S. Sommer. "The Biodiversity of Madagascar: One of The World's Hottest Hotspots on Its Way Out." *Oryx* 35, no. 4 (2001): 346–48.

Ganzhorn, J. U., L. Wilmé, and J. L. Mercier. "Explaining Madagascar's Biodiversity." In *Conservation and Environmental Management in Madagascar*, ed. I. R. Scales, 41–67. London: Routledge, 2014.

Gayle, D., and M. Taylor. "Extinction Rebellion Activists Arrested at Bank of England Protest." *Guardian*, October 14, 2019, https://www.theguardian.com/environment /2019/oct/14/extinction-rebellion-activists-stage-protest-at-bank-of-england.

Gifford, L. "'You Can't Value What You Can't Measure': A Critical Look at Forest Carbon Accounting." *Climatic Change* 161, no. 2 (2020): 291–306.

Gilli, E. "Volcanism-Induced Karst Landforms and Speleogenesis, in the Ankarana Plateau (Madagascar): Hypothesis and Preliminary Research." *International Journal of Speleology* 43, no. 3 (2014): 283–93, http://dx.doi.org/10.5038/1827-806X.43.3.5.

Goedefroit, S., and J. Lombard. *Andolo: L'art funéraire sakalava à Madagascar*. Paris: IRD, 2007.

Goodman, S. "Biological Research Conducted in the General Andasibe Region of Madagascar with Emphasis on Enumerating the Local Biotic Diversity." *Malagasy Nature* 3 (2010): 14–34.

Goodman, S. M., and W. L. Jungers. *Extinct Madagascar: Picturing the Island's Past.* Chicago: University of Chicago Press, 2021.

Goodward, J., and A. Kelly. "Bottom Line on Offsets." World Resources Institute, August 1, 2010, https://www.wri.org/research/bottom-line-offsets.

Greenfield, P. "Is a Madagascan Mine the First to Offset Its Destruction of Rainforest?" *Guardian*, March 9, 2022, https://www.theguardian.com/environment /2022/mar/09/ambatovy-the-madagascan-mine-that-might-prove-carbon-off setting-works-aoe.

Grenier, C. "Genre de vie vezo, pêche 'traditionnelle' et mondialisation sur le littoral sud-ouest de Madagascar." *Annales de géographie* 693 (2013): 549–71.

Grinand, C., F. Rakotomalala, V. Gond, R. Vaudry, M. Bernoux, and G. Vieilledent. "Estimating Deforestation in Tropical Humid and Dry Forests in Madagascar from 2000 to 2010 Using Multi-Date Landsat Satellite Images and the Random Forests Classifier." *Remote Sensing of Environment* 139 (2013): 68–80, https://doi .org/10.1016/j.rse.2013.07.008.

GRiSP (Global Rice Science Partnership). *Rice Almanac*, 4th ed. Los Baños, Philippines: International Rice Research Institute, 2013, http://books.irri.org/9789712 203008_content.pdf.

Grove, K., and J. Pugh. "Assemblage Thinking and Participatory Development: Potentiality, Ethics, Biopolitics." *Geography Compass* 9, no. 1(2015): 1–13.

Grove, R., and R. H. Grove. *Green Imperialism: Colonial Expansion, Tropical Island Edens and the Origins of Environmentalism, 1600–1860*. Cambridge: Cambridge University Press, 1996.

Guardian. "Rabbi, 77, Arrested at Extinction Rebellion's Bank of England Protest— Video Report," October, 14, 2019, https://www.theguardian.com/environment /video/2019/oct/14/rabbi-77-arrested-at-extinction-rebellions-bank-of-england -protest-video-report.

Gudynas, E. "Buen Vivir: Today's Tomorrow." *Development* 54, no. 4 (2011): 441–47.

Hajdu, F., and K. Fischer. "Problems, Causes and Solutions in the Forest Carbon Discourse: A Framework for Analysing Degradation Narratives." *Climate and Development* 9, no. 6 (2017): 537–47.

Hanson, P. W. "Engaging Green Governmentality Through Ritual: The Case of Madagascar's Ranomafana National Park." *Études Océan Indien* 42–43 (2009): 85–113.

Haraway, D. "Situated Knowledges: The Science Question in Feminism and the Privilege of Partial Perspective." *Feminist Studies* 14, no. 3 (1988): 575–99.

Harcourt, W., and I. L. Nelson, eds. *Practising Feminist Political Ecologies: Moving Beyond the "Green Economy."* London: Bloomsbury, 2015.

Harper, G. J., M. K. Steininger, C. J. Tucker, D. Juhn, and F. Hawkins. "Fifty Years of Deforestation and Forest Fragmentation in Madagascar." *Environmental Conservation* 34, no. 4 (2007): 325–33.

Harper, J. "Memories of Ancestry in the Forests of Madagascar." In *Landscape, Memory and History*, ed. P. J. Stewart and A. Strathern, 89–107. London: Pluto Press, 2003.

Hassell, J., M. Roser, E. Ortiz-Ospina, and P. Arriagada. "Global Extreme Poverty." OurWorldInData, n.d., https://ourworldindata.org/poverty.

Hayden, C. *When Nature Goes Public: The Making and Unmaking of Bioprospecting in Mexico*, vol. 1. Princeton: Princeton University Press, 2003.

Henkels, D. "A Close Up of Malagasy Environmental Law." *Vermont Journal of Environmental Law* 3, no. 47 (2001).

Hewitt, A. "Madagascar." In *Structural Adjustment and the African Farmer*, ed. A. Duncan and J. Howell. London: Overseas Development Institute, 1992.

Heynen, N., J. McCarthy, S. Prudham, P. Robbins, eds. *Neoliberal Environments: False Promises and Unnatural Consequences*. London: Routledge, 2007.

Hooper, J. *Feeding Globalization: Madagascar and the Provisioning Trade, 1600–1800*. Athens: Ohio University Press, 2017.

Hornac, J. "Le deboisement et la politique forestiére à Madagascar: Mémoire de stage." In *Mémoires de l'ecole coloniale ENFOM*. Aix-en-Provence: Archives Nationales de France, Archives D'Outre Mer.

Horning, N. R. "Strong Support for Weak Performance: Donor Competition in Madagascar." *African Affairs* 107, no. 428 (2008): 405–31.

Huff, A. "Frictitious Commodities: Virtuality, Virtue and Value in the Carbon Economy of Repair." *Environment and Planning E: Nature and Space* (2021), https://doi.org/10.1177/25148486211015056.

Huff, A., and Y. Orengo. "Resource Warfare, Pacification and the Spectacle of 'Green' Development: Logics of Violence in Engineering Extraction in Southern Madagascar." *Political Geography* 81 (2020): 102195.

Hufty, M., and F. Muttenzer. "Devoted Friends: The Implementation of the Convention on Biological Diversity in Madagascar." In *Governing Global Biodiversity*, ed. P. Le Prestre. Aldershot: Ashgate, 2002.

Igoe, J. *The Nature of the Spectacle: On Images, Money, and Conserving Capitalism*. Tucson: University of Arizona Press, 2017.

IMF. *Republic of Madagascar: Poverty Reduction Strategy Paper Annual Progress Report*. Washington, D.C.: IMF, 2005.

Inverardi-Ferri, C. "The Enclosure of 'Waste Land': Rethinking Informality and Dispossession." *Transactions of the Institute of British Geographers* 43, no. 2 (2018): 230–44.

IUCN. *Manual for the Creation of Blue Carbon Projects in Europe and the Mediterranean*, ed. M. Otero. N.p.: IUCN, 2021, https://www.iucn.org/resources/file/manual-creation-blue-carbon-projects-europe-and-mediterranean.

Jacob, C., J. W. van Bochove, S. Livingstone, T. White, J. Pilgrim, and L. Bennun. "Marine Biodiversity Offsets: Pragmatic Approaches Toward Better Conservation Outcomes." *Conservation Letters* 13, no. 3 (2020): e12711.

Jarosz, L. "Defining and Explaining Tropical Deforestation: Shifting Cultivation and Population Growth in Colonial Madagascar (1896–1940)." *Economic Geography* 69, no. 4 (1993): 366–79.

Jin, C. "Fungi Can Help Concrete Heal Its Own Cracks." *Conversation*, January 19, 2018, https://theconversation.com/fungi-can-help-concrete-heal-its-own-cracks-90375?xid=PS_smithsonian.

Jones, J. P., O. S. Rakotonarivo, and J. H. Razafimanahaka. "Forest Conservation in Madagascar: Past, Present, and Future." In *The New Natural History of Madagascar*, ed. S. Goodman. Princeton, N.J.: Princeton University Press, 2021.

Jones, T. G., L. Glass, S. Gandhi, L. Ravaoarinorotsihoarana, A. Carro, L. Benson, H. R. Ratsimba, et al. "Madagascar's Mangroves: Quantifying Nation-Wide and Ecosystem Specific Dynamics, and Detailed Contemporary Mapping of Distinct Ecosystems." *Remote Sensing* 8, no. 2 (2016): 106.

Kallis, G. "In Defence of Degrowth." *Ecological Economics* 70, no. 5 (2011): 873–880.

Kansang, M. M., and I. Luginaah. "Agrarian Livelihoods Under Siege: Carbon Forestry, Tenure Constraints and the Rise of Capitalist Forest Enclosures in Ghana." *World Development* 113 (2019): 131–42.

Katz, C. "Private Productions of Space and the 'Preservation' of Nature." In *Remaking Reality: Nature at the Millenium*, ed. B. Braun, and N. Castree, 48. London: Routledge, 1998.

Katz, C. "Whose Nature, Whose Culture? Private Productions of Space and the Preservation of Nature." In *Remaking Reality: Nature at the Millennium*, ed. B. Braun and C. Castree, 46–63. London: Routledge, 1998.

Kaufmann, J. C. "Introduction: Recoloring the Red Island." *Ethnohistory* 48, no. 1 (2001): 3–11.

Keck, A., N. P. Sharma, and G. Feder. *Population Growth, Shifting Cultivation, and Unsustainable Agricultural Development: A Case Study in Madagascar*. Washington, D.C.: World Bank Publications, 1994.

Keller, E. "The Banana Plant and the Moon: Conservation and the Malagasy Ethos of Life in Masoala, Madagascar." *American Ethnologist* 35, no. 4 (2008): 650–64.

Keller, E. *Beyond the Lens of Conservation: Malagasy and Swiss Imaginations of One Another*, vol. 20. New York: Berghahn Books, 2015.

Kent, R. K. *Early Kingdoms in Madagascar, 1500–1700*. New York: Holt, Rinehart and Winston, 1970.

Kingston, D. "Bioprospecting for Biodiversity Conservation in Madagascar." REEIS, 2013, https://reeis.usda.gov/web/crisprojectpages/0215326-biodiversity-conservation-and-drug-discovery-in-madagascar.html.

Kingston, D. "Successful Drug Discovery from Natural Products: Methods and Results." Paper read at the Proceedings of the 11th NAPRECA Symposium, at Antananarivo, Madagascar, 2006.

Klein, B. I. "Dina, Domination, and Resistance: Indigenous Institutions, Local Politics, and Resource Governance in Madagascar." *Journal of Peasant Studies* (2023): 1–30.

Klein, J., B. Réau, and M. Edwards. "Zebu Landscapes: Conservation and Cattle in Madagascar." In *Greening the Great Red Island: Madagascar in Nature and Culture*, ed. J. Kaufmann, 157–78. Africa Institute of South Africa, 2008.

Klein, N. "How Power Profits from Disaster." *Guardian,* July 6, 2017, https://www.the guardian.com/us-news/2017/jul/06/naomi-klein-how-power-profits-from-disaster.

Klein, N. *The Shock Doctrine: The Rise of Disaster Capitalism.* New York: Macmillan, 2007.

Klinger, J. M. *Rare Earth Frontiers: From Terrestrial Subsoils to Lunar Landscapes.* Ithaca, N.Y.: Cornell University Press, 2018.

Koehn, F. E., and G. T. Carter. "The Evolving Role of Natural Products in Drug Discovery." *Nature Reviews Drug Discovery* 4, no. 3 (2005): 206–20.

Kopnina, H. "Half the Earth for People (or More)? Addressing Ethical Questions in Conservation." *Biological Conservation* 203 (2016): 176–85.

Kuletz, V. *The Tainted Desert: Environmental and Social Ruin in the American West.* New York: Routledge, 2016.

Kull, C. A. *Isle of Fire: The Political Ecology of Landscape Burning in Madagascar.* Chicago: University of Chicago Press, 2004.

Laird, S. *Access and Benefit-Sharing: Key Points for Policy-Makers, the Pharmaceutical Industry,* The ABS Capacity Development Initiative, GIZ. Cape Town: University of Cape Town and People and Plants International, 2015.

Laird, S., and R. Wynberg. *Access and Benefit-Sharing in Practice: Trends in Partnerships Across Sectors.* Montreal: CBD Technical Series/UNEP, 2008.

Laird, S. A. *Biodiversity and Traditional Knowledge: Equitable Partnerships in Practice.* N.p.: Routledge, 2010.

Lansing, D. M. "Performing Carbon's Materiality: The Production of Carbon Offsets and the Framing of Exchange." *Environment and Planning A* 44, no. 1 (2012): 204–20.

Lansing, D. M. "Understanding Smallholder Participation in Payments for Ecosystem Services: The Case of Costa Rica." *Human Ecology* 45, no. 1 (2017): 77–87.

Larson, P. M. *History and Memory in the Age of Enslavement: Becoming Merina in Highland Madagascar, 1770–1822.* Portsmouth, N.H.: Heinemann, 2000.

Lavauden, L. "Histoire de la législation et de l'administration forestière à Madagascar." *Revue des Eaux et Forêts* 72 (1934): 949–60.

Lave, R., and M. Robertson. "Biodiversity Offsetting." In *The Routledge Handbook of the Political Economy of Science,* ed. D. Tyfield et al., 224–36. Oxon, UK: Routledge, 2017.

Le Billon, P. "Crisis Conservation and Green Extraction: Biodiversity Offsets as Spaces of Double Exception." *Journal of Political Ecology,* 28, no. 1: 854–88.

Lee, C. "Griffithsin, a Highly Potent Broad-Spectrum Antiviral Lectin from Red Algae: From Discovery to Clinical Application." *Marine Drugs* 17, no. 10 (2019): 567.

Lerner, S. *Sacrifice Zones: The Front Lines of Toxic Chemical Exposure in the United States.* Cambridge, Mass.: MIT Press, 2010.

L'Express de Madagascar. "Un jardin d'essais cultural a Namisana." December 28, 2016, https://lexpress.mg/28/12/2016/un-jardin-dessais-culturaux-a-nanisana/.

Li, T. M. *The Will to Improve: Governmentality, Development, and the Practice of Politics.* Durham, N.C.: Duke University Press, 2007.

Lomas, C. "Mining the Moon: Earth's Back-Up Plan?" *Deutsche Welle*, February 2, 2018, https://www.dw.com/en/mining-the-moon-earths-back-up-plan/a-422 86299.

Lovell, H. "Climate Change, Markets and Standards: The Case of Financial Accounting." *Economy and Society* 43, no. 2 (2014): 260–84, https://doi.org/10.1080/0308 5147.2013.812830.

Lovell, H., H. Bulkeley, and D. Liverman. "Carbon Offsetting: Sustaining Consumption?" *Environment and Planning A* 41, no. 10 (2009): 2357–79.

Lovell, H., and N. S. Ghaleigh. "Climate Change and the Professions: The Unexpected Places and Spaces of Carbon Markets." *Transactions of the Institute of British Geographers* 38, no. 3 (2013): 512–16.

Lovell, H., and D. Liverman. "Understanding Carbon Offset Technologies." *New Political Economy* 15, no. 2 (2010): 255–73.

MacDonald, K. "The Devil Is in the (Bio)diversity: Private Sector 'Engagement' and the Restructuring of Biodiversity Conservation." *Antipode* 42, no. 3 (2010): 513–50.

Macilwain, C. "When Rhetoric Hits Reality in Debate on Bioprospecting." *Nature* 392, no. 6676 (1998): 535.

MacKenzie, D. "Finding the Ratchet: The Political Economy of Carbon Trading." *Post Autistic Economics Review* 42 (2007): 8–17.

MacKenzie, D. "Is Economics Performative? Option Theory and the Construction of Derivatives Markets." *Journal of the History of Economic Thought* 28, no. 1 (2006): 29–55.

Mahanty, S. *Unsettled Frontiers: Market Formation in the Cambodia-Vietnam Borderlands*. Ithaca: N.Y.: Cornell University Press, 2022.

Mahanty, S., and B. Neimark. "The Green Gig Economy: Precarious Workers Are on the Frontline of Climate Change Fight." *Conversation*, April 21, 2020, https://the conversation.com/the-green-gig-economy-precarious-workers-are-on-the-front line-of-climate-change-fight-133392.

Marcus, R. R., and C. A. Kull. "The Politics of Conservation in Madagascar." *African Studies Quarterly* 3, no. 2 (1999): 1–8.

Marcus, R. R., and A. M. Ratsimbaharison. "Political Parties in Madagascar: Neopatrimonial Tools or Democratic Instruments?" *Party Politics* 11, no. 4 (2005): 495–512.

Marley, B. J. "The Coal Crisis in Appalachia: Agrarian Transformation, Commodity Frontiers and the Geographies of Capital." *Journal of Agrarian Change* 16, no. 2 (2016): 225–54.

Martin, A., B. Coolsaet, E. Corbera, N. M. Dawson, J. A. Fraser, J. A., I. Lehmann, and I. Rodriguez. "Justice and Conservation: The Need to Incorporate Recognition." *Biological Conservation* 197 (2016): 254–61.

Martinez-Alier, J. *The Environmentalism of the Poor: A Study of Ecological Conflicts and Valuation*. Northampton, Mass.: Edward Elgar, 2002.

Marx, K. *Capital. A Critique of Political Economy*, vol. 1. London: Penguin Books, 1976.

McAfee, K. "Selling Nature to Save It? Biodiversity and Green Developmentalism." *Environment and Planning D: Society and Space* 17, no. 2 (1999): 133–54.

McAfee, K., and E. N. Shapiro. "Payments for Ecosystem Services in Mexico: Nature, Neoliberalism, Social Movements, and the State." *Annals of the Association of American Geographers* 100, no. 3 (2010): 579–99.

McCourt, B. *BGS International Activities* (2010), http://nora.nerc.ac.uk/id/eprint /9191/.

McKibben, B. "Citing Climate Change, BlackRock Will Start Moving Away from Fossil Fuels." *New Yorker*, January 16, 2020, https://www.newyorker.com/news/daily -comment/citing-climate-change-blackrock-will-start-moving-away-from-fossil -fuels?verso=true.

McLeod, E., G. L. Chmura, S. Bouillon, R. Salm, M. Björk, C. M. Duarte, C. Lovelock, et al. "A Blueprint for Blue Carbon: Toward an Improved Understanding of the Role of Vegetated Coastal Habitats in Sequestering CO_2." *Frontiers in the Ecology and the Environment* 9, no. 10 (2011): 552–60, https://doi.org/10.1890/110004.

Merino, N., H. S. Aronson, D. P. Bojanova, J. Feyhl-Buska, M. L. Wong, S. Zhang, and D. Giovannelli. "Living at the Extremes: Extremophiles and the Limits of Life in a Planetary Context." *Frontiers in Microbiology* 10 (2019): 780.

Metz, H. *Indian Ocean: Five Island Countries*. Washington, D.C.: Federal Research Division, 1995.

Mignolo, W. D. "Delinking: The Rhetoric of Modernity, the Logic of Coloniality and the Grammar of De-coloniality." *Cultural Studies* 21, nos. 2–3 (2007): 449–514.

Mikulewicz, M., L. Johnson, M. Mills-Novoa and B. Neimark. "Understanding Adaptation as Labor: Implications for Critical Climate Research." Paper presentation at the American Association of Geographers (AAG) Annual Meeting, Denver, US, March 25, 2023. https://aag.secureplatform.com/aag2023/solicitations/39 /sessiongallery/6604.

Miller, J., C. Birkinshaw, and M. Callmander. "The Madagascar International Cooperative Biodiversity Group (ICBG): Using Natural Products Research to Build Science Capacity." *Ethnobotany Research and Applications* 3 (2005): 283–86.

Miller, J. S. "The Discovery of Medicines from Plants: A Current Biological Perspective." *Economic Botany* 65, no. 4 (2011): 396–407.

Miller, J. S. "Impact of the Convention on Biological Diversity: The Lessons of Ten Years of Experience with Models for Equitable Sharing of Benefits." In *Biodiversity and the Law: Intellectual Property, Biotechnology and Traditional Knowledge*, ed. C. McManis. London: Earthscan, 2007.

Milne, S. "Grounding Forest Carbon: Property Relations and Avoided Deforestation in Cambodia." *Human Ecology* 40, no. 5 (2012): 693–706.

Milne, S., and B. Adams. "Market Masquerades: Uncovering the Politics of Community-Level Payments for Environmental Services in Cambodia." *Development and Change* 43, no. 1 (2012): 133–58.

Ministère de L'Environnement, des Forêts, et du Tourisme (MEFT). *Manuel de création des Aires Marine Protégées a Madagascar*. Antannarivo: DGEF, 2009.

Mittermeier, R. A., N. Myers, J. B. Thomsen, G. A. Da Fonseca, and S. Olivieri. "Biodiversity Hotspots and Major Tropical Wilderness Areas: Approaches to Setting Conservation Priorities." *Conservation Biology* 12, no. 3 (1998): 516–20.

Mittermeier, R. A., G. P. Robles, M. Hoffman, J. Pilgrim, T. Brooks, and C. G. Mittermeier. *Hotspots Revisited*. Mexico City: Cemex, 2004.

Mollett, S., and T. Kepe. *Land Rights, Biodiversity Conservation and Justice: Rethinking Parks and People*. Oxon, UK: Routledge, 2018.

Montagne, P., and A. Bertrand. *Histoire des politiques forestières au Niger, au Mali et à Madagascar*. Paris: L'Harmattan, 2006.

Montagne, P., and B. Ramamonjisoa. "Politiques forestières à Madagascar entre répression et autonomie des acteurs." *Économie Rurale: Agricultures, Alimentations, Territoires* 4–5, nos. 294–95 (2006): 9–26.

Moore, J. "Industrial Revolution II." jasonwmoore.wordpress.com, July 4, 2013, https://jasonwmoore.wordpress.com/tag/industrial-revolution/.

Moore, J. W. *Capitalism in the Web of Life: Ecology and the Accumulation of Capital*. London: Verso Books, 2015.

Moore, J. W. "The Capitalocene Part II: Accumulation by Appropriation and the Centrality of Unpaid Work/Energy." *Journal of Peasant Studies* 45, no. 2 (2018): 237–79.

Moore, J. W. "Nature, Geopower, and Capitalogenic Appropriation." jasonwmoore .wordpress.com, November 8, 2016, https://jasonwmoore.wordpress.com/2016 /11/08/nature-geopower-capitalogenic-appropriation/.

Moore, J. W. "Sugar and the Expansion of the Early Modern World-Economy: Commodity Frontiers, Ecological Transformation, and Industrialization." *Review (Fernand Braudel Center)* (2000): 409–33.

Moser, C., C. Barrett, and B. Minten. "Spatial Integration at Multiple Scales: Rice Markets in Madagascar." *Agricultural Economics* 40, no. 3 (2009): 281–94.

Müller, F., T. Johanna, and T. Kalt. "Hydrogen Justice." *Environmental Research Letters* 17, no. 11 (2022): 115006.

Myers, N. "The Biodiversity Challenge: Expanded Hot-Spots Analysis." *Environmentalist* 10, no. 4 (1990): 243–56.

Myers, N., R. A. Mittermeier, C. G. Mittermeier, G. A. Da Fonseca, and J. Kent. "Biodiversity Hotspots for Conservation Priorities." *Nature* 403, no. 6772 (2000): 853–58.

Neimark, B. "Bioprospecting and Biopiracy." In *The International Encyclopedia of Geography: People, the Earth, Environment and Technology*, ed. D. Richardson, N. Castree, M. F. Goodchild, A. L. Kobayashi, W. Liu, and R. A. Marston. Chichester, UK: John Wiley & Sons, 2017.

Neimark, B. "Green Grabbing at the 'Pharm' Gate: Rosy Periwinkle Production in Southern Madagascar." *Journal of Peasant Studies* 39, no. 2 (2012): 423–45.

Neimark, B. *Industrial Heartlands of Nature: The Political Economy of Biological Prospecting in Madagascar*. New Brunswick, N.J.: Rutgers, 2009.

Neimark, B., and L. Tilghman. "Bioprospecting a Biodiversity Hotspot: The Political Economy of Natural Products Drug Discovery for Conservation Goals in Madagascar." In *Conservation and Environmental Management in Madagascar*, ed. I. R. Scales, 295–322. London: Routledge, 2014.

Neimark, B., S. Mahanty, and W. Dressler. "Mapping Value in a 'Green' Commodity Frontier: Revisiting Commodity Chain Analysis." *Development and Change* 47, no. 2 (2016): 240–65.

Neimark, B., S. Mahanty, W. Dressler, and C. Hicks. "Not Just Participation: The Rise of the Eco-precariat in the Green Economy." *Antipode* 52, no. 2 (2020): 496–521.

Neimark, B., S. Osterhoudt, H. Alter, and A. Gradinar. "A New Sustainability Model for Measuring Changes in Power and Access in Global Commodity Chains: Through a Smallholder Lens." *Palgrave Communications* 5, no. 1 (2019), https://doi.org/10.1057/s41599-018-0199-0.

Neimark, B., S. Osterhoudt, L. Blum, and T. Healy. "Mob Justice and 'The Civilized Commodity.'" *Journal of Peasant Studies* 48, no. 4 (2021): 734–53, https://doi.org/10.1080/03066150.2019.1680543.

Neimark, B. D. "Biofuel Imaginaries: The Emerging Politics Surrounding 'Inclusive' Private Sector Development in Madagascar." *Journal of Rural Studies* 45 (2016): 146–56.

Neimark, B. D., and R. A. Schroeder. "Hotspot Discourse in Africa: Making Space for Bioprospecting in Madagascar." *African Geographical Review* 28, no. 1 (2009): 43–69.

Neimark, B. D., and S. Vermeylen. "A Human Right to Science? Precarious Labor and Basic Rights in Science and Bioprospecting." *Annals of the American Association of Geographers* 107, no. 1 (2017): 167–82.

Neimark, B. D., and B. Wilson. "Re-mining the Collections: From Bioprospecting to Biodiversity Offsetting in Madagascar." *Geoforum* 66 (2015): 1–10.

New York Times. "BlackRock C.E.O. Larry Fink: Climate Crisis Will Reshape Finance." February 24, 2020, https://www.nytimes.com/2020/01/14/business/dealbook/larry-fink-blackrock-climate-change.html.

New York Times. "Spider Silk Is Stronger Than Steel. It Also Assembles Itself." November 4, 2020, https://www.nytimes.com/2020/11/04/science/spider-silk-web-self-assembly.html.

New York Times. "Turning Air Into Perfume." December 12, 2021, https://www.nytimes.com/2021/12/16/fashion/air-eau-de-parfum-perfume.html.

Nightingale, A. J. "A Feminist in the Forest: Situated Knowledges and Mixing Methods in Natural Resource Management." *ACME: An International Journal for Critical Geographies* 2, no. 1 (2003): 77–90.

NOAA. "Coastal Blue Carbon." https://oceanservice.noaa.gov/podcast/may14/mw124-bluecarbon.html.

Odling-Smee, L. "Conservation: Dollars and Sense." *Nature* 437, no. 7059 (2005): 614–17, 614.

Offset Guide. "How to Acquire Carbon Offset Credits." https://www.offsetguide.org/understanding-carbon-offsets/how-to-acquire-carbon-offset-credits/#_ftnref1.

Oliver, T. A., K. L. Oleson, H. Ratsimbazafy, D. Raberinary, S. Benbow, and A. Harris. "Positive Catch and Economic Benefits of Periodic Octopus Fishery Closures: Do

Effective, Narrowly Targeted Actions 'Catalyze' Broader Management?" *PLoS One* 10, no. 6 (2015): e0129075.

Olson, S. "The Robe of the Ancestors: Forests in the History of Madagascar." *Journal of Forest History* 28, no. 4 (1984): 174–86, 178.

Osborne, T., and E. Shapiro-Garza. "Embedding Carbon Markets: Complicating Commodification of Ecosystem Services in Mexico's Forests." *Annals of the American Association of Geographers* 108, no. 1 (2018): 88–105.

Osterhoudt, S., S. S. Galvin, D. J. Graef, A. K. Saxena, and M. R. Dove. "Chains of Meaning: Crops, Commodities, and the 'In-Between' Spaces of Trade." *World Development* 135 (2020): 105070.

Osterhoudt, S. R. *Vanilla Landscapes: Meaning, Memory, and the Cultivation of Place in Madagascar.* New York: New York Botanical Garden, 2017.

Ouma, S., L. Johnson, and P. Bigger. "Rethinking the Financialization of 'Nature.'" *Environment and Planning A: Economy and Space* 50, no. 3 (2018): 500–11.

Paprocki, K. "The Climate Change of Your Desires: Climate Migration and Imaginaries of Urban and Rural Climate Futures." *Environment and Planning D: Society and Space* 38, no. 2 (2020): 248–66.

Parenti, C. "Environment-Making in the Capitalocene." In *Anthropocene or Capitalocene? Nature, History, and the Crisis of Capitalism*, ed. J. Moore, 166–83. Oakland, Calif: PM Press, 2016.

Parry, B. *Trading the Genome: Investigating the Commodification of Bio-information.* New York: Columbia University Press, 2004.

Peet, R., and M. Watts. *Liberation Ecologies: Environment, Development and Social Movements.* London: Routledge, 1996.

Peluso, N. L., and M. Watts, eds. *Violent Environments.* Ithaca, N.Y.: Cornell University Press, 2001.

Pendleton, L., D. C. Donato, B. C. Murray, S. Crooks, W. A. Jenkins, S. Sifleet, C. Craft, et al. "Estimating Global 'Blue Carbon' Emissions from Conversion and Degradation of Vegetated Coastal Ecosystems." *PLoS One* 7, no. 9 (2012), https://doi.org/10.1371/journal.pone.0043542.

Perelman, M. "Exclusive: Madagascar's President Defends Controversial Homegrown COVID-19 Cure." *France 24*, December 5, 2020, https://www.france24.com/en/africa/20200512-exclusive-madagascar-s-president-defends-controversial-homegrown-COVID-19-cure.

Petkova, E., A. Larson, and P. Pacheco. "Forest Governance, Decentralization and REDD+ in Latin America." *Forests* 1, no. 4 (2010): 250–54.

Phelps, J., E. L. Webb, and A. Agrawal. "Does REDD+ Threaten to Recentralize Forest Governance?" *Science* 328, no. 5976 (2010): 312–13.

Pollini, J. "Slash-and-Burn Cultivation and Deforestation in the Malagasy Rain Forests: Representations and Results." PhD diss., Cornell University, 2007.

Pollini, J., N. Hockley, and F. D. Muttenzer. "The Transfer of Natural Resource Management Rights to Local Communities." In *Conservation and Environmental Management in Madagascar*, ed. I. Scales, 196–216. London: Routledge, 2014.

Poudyal, M., B. S. Ramamonjisoa, N. Hockley, O. S. Rakotonarivo, J. M. Gibbons, R. Mandimbiniaina, A. Rasoamanana, and J. P. Jones. "Can REDD+ Social Safeguards Reach the 'Right' People? Lessons from Madagascar." *Global Environmental Change* 37 (2016): 31–42.

Prieto, M. "Indigenous Resurgence, Identity Politics, and the Anticommodification of Nature: The Chilean Water Market and the Atacameno People." *Annals of the American Association of Geographers* 112, no. 2 (2022): 487–504.

Raik, D. "Forest Management in Madagascar: An Historical Overview." *Madagascar Conservation and Development* 2, no. 1 (2007).

Rakotonjatovo, B. H., A. Rasolofoarimanga, H. Andriamanantoanina, L. Ranarivelo, J. Maharavo, L. Ramaroson, and M. Ratsimbason. "A Microfluorimetric Method to Screen Marine Products for Antimalarial Activity: Preliminary Results." In *Proceedings of the 11th NAPRECA Symposium*, 154–60. Antananarivo, 2006.

Ramanantsoavina, G. *Histoire de la politique forestière à Madagascar*. Antananarivo: DGEF, 1963.

Ramanantsoavina, G., and A. Rakotomanampison. *Fertilisation des plantations industrielles de pins à Madagascar*. Antananarive: GERDAT-CTFT, 1973.

Randriamanamisa, R., and G. J. Stads. *Madagascar: Recent Developments in Public Agricultural Research*, Agricultural Science and Technology Indicators. Rome: Food and Agriculture Organization of the United Nations, 2010.

Randrianja, S., and S. Ellis. *Madagascar: A Short History*. London: C. Hurst, 2009.

Ratsifandrihamanana, N. "Community Management of Natural Resources: The Future of Madagascar." WWF, January 10, 2018, https://wwf.panda.org/?320350 /COURRIER2DDES2DLECTEURS2Dpar2DNanie2DRatsifandrihamanana.

Ratsimbaharison, A. M. "The Obstacles and Challenges to Democratic Consolidation in Madagascar (1992–2009)." *International Journal of Political Science* 2, no. 2 (2016): 1–17.

Ratsimbason, M., L. Ranarivelo, H. R. Juliani, and J. E. Simon. "Antiplasmodial Activity of Twenty Essential Oils from Malagasy Aromatic Plants." In *African Natural Plant Products: New Discoveries and Challenges in Chemistry and Quality*, ed. H. R. Juliani, J. E. Simon, and C. T. Ho, 209–15. Washington, D.C.: American Chemical Society, 2009.

Reid, W. *Biodiversity Prospecting: Using Genetic Resources for Sustainable Development*. Washington D.C.: WRI, 1993.

Ribot, J. C. "Theorizing Access: Forest Profits along Senegal's Charcoal Commodity Chain." *Development and Change* 29, no. 2 (1998): 307–41.

Riofrancos, T. *Resource Radicals: From Petro-nationalism to Post-extractivism in Ecuador*. Durham: Duke University Press, 2020.

Robbins, P. *Political Ecology: A Critical Introduction*, 3rd ed. Hoboken, N.J.: John Wiley & Sons, 2019.

Robertson, M. "Measurement and Alienation: Making a World of Ecosystem Services." *Transactions of the Institute of British Geographers* 37, no. 3 (2012), 386–401, 3.

Robinson, D. *Biodiversity, Access and Benefit-Sharing: Global Case Studies.* Oxon, UK: Routledge, 2014, http://dx.doi.org/10.4324/9781315882819.

Rocheleau, D. E., L. Thomas-Slayter, and E. Wangari. *Feminist Political Ecology: Global Perspectives and Local Experience.* London: Routledge, 1997.

Rodney, W. *How Europe Underdeveloped Africa.* London: Verso Trade, 2018.

Roederer, P., and F. Bourgeat. "Carte des sols de Madagascar au 14000000è." In *Atlas de Madagascar.* Antananarivo: Université de Madagascar, 1971.

Rosenthal, J., F. Katz, and A. Bull. *Microbial Diversity and Bioprospecting.* Washington, D.C.: ASM Press, 2004.

Sarrasin, B. "The Mining Industry and the Regulatory Framework in Madagascar: Some Developmental and Environmental Issues." *Journal of Cleaner Production* 14, nos. 3–4 (2006): 388–96.

Scales, I., D. Friess, L. Glass, and L. Ravaoarinorotsihoarana. "Rural Livelihoods and Mangrove Degradation in South-West Madagascar: Lime Production as an Emerging Threat." *Oryx* 52, no. 4 (2017), https://doi.org/10.1017/S0030605316001630.

Scales, I. R. "The Future of Conservation and Development in Madagascar: Time for a New Paradigm?" *Madagascar Conservation and Development* 9, no. 1 (2014): 5–12.

Scales, I. R. "Paying for Nature: What Every Conservationist Should Know About Political Economy." *Oryx* 49, no. 2 (2015): 226–31.

Schiebinger, L. *Plants and Empire: Colonial Bioprospecting in the Atlantic World.* Harvard University Press, 2007.

Schiebinger, L., and C. Swan, eds. *Colonial Botany: Science, Commerce, and Politics in the Early Modern World.* Philadelphia: University of Pennsylvania Press, 2007.

Schleicher, J., J. G. Zaehringer, C. Fastré, B. Vira, P. Visconti, and C. Sandbrook. "Protecting Half of the Planet Could Directly Affect over One Billion People." *Nature Sustainability* 2, no. 12 (2019): 1094–96.

Schroeder, R. *Shady Practices: Agroforestry and Gender Politics in the Gambia.* Berkeley: University of California Press, 1999.

Schroeder, R. A. "Community, Forestry and Conditionality in the Gambia." *Africa* 69, no. 1 (1999): 1–22.

Schroeder, R. A., and R. P. Neumann. "Manifest Ecological Destinies: Local Rights and Global Environmental Agendas." *Antipode* 27, no. 4 (1995): 321–24.

Scott, J. C. "Seeing Like a State." In *Seeing Like a State: How Certain Schemes to Improve the Human Condition Have Failed.* New Haven: Yale University Press, 1998.

Seagle, C. "Inverting the Impacts: Mining, Conservation and Sustainability Claims near the Rio Tinto/QMM Ilmenite Mine in Southeast Madagascar." *Journal of Peasant Studies* 39, no. 2 (2012): 447–77.

Setyowati, A. B. "Governing the Ungovernable: Contesting and Reworking REDD+ in Indonesia." *Journal of Political Ecology* 27, no. 1 (2020): 456–75.

Seymour, F., and J. Busch. *Why Forests? Why Now? The Science, Economics, and Politics of Tropical Forests and Climate Change.* Washington, D.C.: Center for Global Development, 2016.

Shankleman, J., and A. Rathi. "Wall Street's Favorite Climate Solution Is Mired in Disagreements." *Bloomberg*, June 1, 2021, https://www.bloomberg.com/news /features/2021-06-02/carbon-offsets-new-100-billion-market-faces-disputes-over -trading-rules.

Sharp, L. A. *The Sacrificed Generation*. Berkeley: University of California Press, 2002.

Shiva, V. *Biopiracy: The Plunder of Knowledge and Nature*. Boston: South End Press, 1997.

Sibanda, P. "The Dimensions of 'Hunhu/Ubuntu'(Humanism in the African sense): The Zimbabwean Conception." *Dimensions* 4, no. 1 (2014): 26–28.

Silver, J. J., N. J. Gray, L. M. Campbell, L. W. Fairbanks, and R. L. Gruby. "Blue Economy and Competing Discourses in International Oceans Governance." *Journal of Environment and Development* 24, no. 2 (2015): 135–60.

Sittenfeld, A., and A. Lovejoy. "Biodiversity Prospecting Frameworks: The INBio Experience in Costa Rica." In *Protection of Global Biodiversity: Converging Strategies*, ed. L. Guruswamy and J. McNeely, 223–44. Durham, N.C.: Duke University Press, 1996.

Smith, N. "Nature as Accumulation Strategy." *Socialist Register* 43 (2007): 33.

Sneader, W. *Drug Discovery: A History*. Hoboken, N.J.: John Wiley & Sons, 2005.

Sodikoff, G. M. *Forest and Labor in Madagascar: From Colonial Concession to Global Biosphere*. Bloomington: Indiana University Press, 2012.

Soejarto, D. D., H. H. S. Fong, G. T. Tan, H. J. Zhang, C. Y. Ma, S. G. Franzblau, and G. R. Dietzman. "Ethnobotany/Ethnopharmacology and Mass Bioprospecting: Issues on Intellectual Property and Benefit-Sharing." *Journal of Ethnopharmacology* 100, nos. 1–2 (2005): 15–22.

Solis, M. "Coronavirus Is the Perfect Disaster for 'Disaster Capitalism.'" *Vice*, March 13, 2020, https://www.vice.com/en_uk/article/5dmqyk/naomi-klein-inter view-on-coronavirus-and-disaster-capitalism-shock-doctrine.

Sovacool, B. K., S. H. Ali, M. Bazilian, B. Radley, B. Nemery, J. Okatz, D. Mulvaney. "Sustainable Minerals and Metals for a Low-Carbon Future." *Science* 367, no. 6473 (2020): 30–33.

Standing, G. *The Precariat: The New Dangerous Class*. London: Bloomsbury Academic, 2011.

Styger, E., J. Rakotoarimanana, R. Rabevohitra, and E. Fernandes. "Indigenous Fruit Trees of Madagascar: Potential Components of Agroforestry Systems to Improve Human Nutrition and Restore Biological Diversity." *Agroforestry Systems* 46, no. 3 (1999): 289–310.

Sullivan, S. "After the Green Rush? Biodiversity Offsets, Uranium Power and the 'Calculus of Casualties' in Greening Growth." *Human Geography* 6, no. 1 (2013): 80–101.

Sullivan, S. "Banking Nature? The Spectacular Financialisation of Environmental Conservation." *Antipode* 45, no. 1 (2013): 198–217.

Sullivan, S. "Nature on the Move III: (Re)Countenancing an Animate Nature." *New Proposals: Journal of Marxism and Interdisciplinary Inquiry* 6, nos. 1–2 (2013): 50–71.

Sultana, F. "Decolonizing Development Education and the Pursuit of Social Justice." *Human Geography* 12, no. 3 (2019): 31–46, 33.

Sultana, F. "Political Ecology 1: From Margins to Center." *Progress in Human Geography* 45, no. 1 (2021): 156–65.

Sundberg, J. "Feminist Political Ecology." In *The International Encyclopedia of Geography: People, the Earth, Environment and Technology*, ed. D. Richardson, 1–12. Wiley-Blackwell, 2016.

Svarstad, H. "A Global Political Ecology of Bioprospecting." In *Political Ecology Across Spaces, Scales and Social Groups*, ed. S. Paulson and L. Gezon, 239–56. New Brunswick, N.J.: Rutgers University Press, 2004.

Ten Kate, K., and S. A. Laird. "Biodiversity and Business: Coming to Terms with the 'Grand Bargain.'" *International Affairs* 76, no. 2 (2000): 241–64.

Ten Kate, K., and S. A. Laird. *The Commercial Use of Biodiversity: Access to Genetic Resources and Benefit-Sharing*. Milton, UK: Routledge, 2019.

Teyssier, A., R. Andrianirina-Ratsialonana, R. Razafindralambo, Y. Razafindrakoto. "Decentralization of Land Management in Madagascar: Process, Innovations, and Observation of the First Outcomes." Paper presented at the Annual World Bank Conference on Land Administration, Washington, D.C., 2008.

Thunberg, G., A. Taylor, et al. "Think We Should Be at School? Today's Climate Strike Is the Biggest Lesson of All." *Guardian*, March 15, 2019, https://www.theguardian.com/commentisfree/2019/mar/15/school-climate-strike-greta-thunberg.

Trefis Team. "With $9.5 Trillion In Assets, Is BlackRock Stock Fairly Priced At $910?" *Forbes*, August 19, 2021, https://www.forbes.com/sites/greatspeculations/2021/08/19/with-95-trillion-in-assets-is-blackrock-stock-fairly-priced-at-910/?sh=431fff895b5b.

Tsing, A. L. *Friction: An Ethnography of Global Connection*. Princeton, N.J.: Princeton University Press, 2011.

Tsing, A. L. "Inside the Economy of Appearances." *Public Culture* 12, no. 1 (2000): 115–44.

UNCTAD. *Foreign Direct Investment in LDCs: Lessons Learned from the Decade 2001–2010 and the Way Forward*. New York: United Nations, 2011.

Vandergeest, P., and N. L. Peluso. "Empires of Forestry: Professional Forestry and State Power in Southeast Asia, Part 2." *Environment and History* 12, no. 4 (2006): 359–93.

Vaughan, A. "No Evidence 'Madagascar Cure' for COVID-19 Works, says WHO," *New Scientist*, May 15, 2020, https://www.newscientist.com/article/2243669-no-evidence-madagascar-cure-for-COVID-19-works-says-who/#ixzz6Oh0NaQVZ.

Verin, P. "Deux étranges statues en chloritoschiste de Madagascar." *Publications de la Société Française D'histoire des Outre-mers* 5, no. 1 (1981): 155–60.

Vian, J. E. "Five Ways 'Green' Carbon Policies Damage Forests—and How We Can Fix the Problem." June 9, 2021, https://theconversation.com/five-ways-green-carbon-policies-damage-forests-and-how-we-can-fix-the-problem-162132?utm_source=twitter&utm_medium=bylinetwitterbutton.

Vieilledent, G., C. Grinand, F. A. Rakotomalala, R. Ranaivosoa, J-R Rakotoarija-ona, T. Allnutt, and F. Achard. "Combining Global Tree Cover Loss Data with Historical National Forest Cover Maps to Look at Six Decades of Deforestation and Forest Fragmentation in Madagascar." *Biological Conservation* 222 (2018): 189–97.

Walsh, A. *Made in Madagascar: Sapphires, Ecotourism, and the Global Bazaar.* Toronto: University of Toronto Press, 2012.

Watts, M. "Resource Curse? Governmentality, Oil and Power in the Niger Delta, Nigeria." *Geopolitics* 9, no. 1 (2004): 50–80.

"Webinar 3: Blue Forests Science for the Oceans We Want," https://us02web.zoom.us/j/82013671252.

Wes, P., and J. Carrier. "Ecotourism and Authenticity: Getting Away from It All?" *Current Anthropology* 45, no. 4 (2004): 483–98.

West, P. *Conservation Is Our Government Now: The Politics of Ecology in Papua New Guinea.* Durham, N.C.: Duke University Press, 2006.

West, P. "Critical Approaches to Dispossession in the Melanesian Pacific: Conservation, Voice, and Collaboration." Presentation delivered at the Second Biennial Conference of the Political Ecology Network (POLLEN), Political Ecology, the Green Economy, and Alternative Sustainabilities, Oslo Metropolitan University, Oslo, Norway, June, 19–22, 2018, https://politicalecologynetwork.org/2018/09/01/paige-west-keynote-lecture-at-pollen18-conference-oslo-20-june-2018/.

Wilmé, L., S. M. Goodman, and J. U. Ganzhorn. "Biogeographic Evolution of Madagascar's Microendemic Biota." *Science* 312, no. 5776 (2006): 1063–65.

Wilson, E. O. *Half-Earth: Our Planet's Fight for Life.* New York: W. W. Norton, 2016.

World Bank. "Madagascar: Balancing Conservation and Exploitation of Fisheries Resources." June 8, 2020, https://www.worldbank.org/en/news/feature/2020/06/08/madagascar-balancing-conservation-and-exploitation-of-fisheries-resources.

World Bank. "What Is the Blue Economy?" June 5, 2016, https://www.worldbank.org/en/news/infographic/2017/06/06/blue-economy.

WWF. "The Bezos Earth Fund & WWF: Investment in Community and Climate." https://www.worldwildlife.org/pages/the-bezos-earth-fund-wwf-investment-in-community-and-climate.

Wynberg, R., and S. Laird. "Bioprospecting: Tracking the Policy Debate." *Environment: Science and Policy for Sustainable Development* 49, no. 10 (2007): 20–32.

Yancho, J. M. M., T. G. Jones, S. R. Gandhi, C. Ferster, A. Lin, and L. Glass. "The Google Earth Engine Mangrove Mapping Methodology (GEEMMM)." *Remote Sensing* 12, no. 22 (2020): 3758.

Yusoff, K. *A Billion Black Anthropocenes or None.* Minneapolis: University of Minnesota Press, 2018.

Zerner, C. "Introduction: Toward a Broader Vision of Justice and Nature Conservation." In *People, Plants, and Justice*, ed. C. Zerner, 3–20. New York: Columbia University Press, 2000.

Zhu, A. L., and B. Klein. "The Rise of Flexible Extraction: Boom-Chasing and Subject-Making in Northern Madagascar." *Geoforum* (2022), https://doi.org/10.1016/j.geoforum.2022.06.005.

Zimmer, C. "Bringing Them Back to Life." *National Geographic* (2013), https://www.nationalgeographic.com/magazine/2013/04/species-revival-bringing-back-extinct-animals/.

Zografos, C., and P. Robbins. "Green Sacrifice Zones, or Why a Green New Deal Cannot Ignore the Cost Shifts of Just Transitions." *One Earth* 3, no. 5 (2020): 543–46.

Index

Note: Locators in *italics* refer to figures.

About the Author

Benjamin Neimark is a senior lecturer at the School of Business and Management, and a Fellow at the Institute of Social Science and Humanities (IHSS) at Queen Mary University of London. He is a human geographer and political ecologist ("political ecology" defined as the intersections of ecology and a broadly defined political economy) whose research focuses on politics of biological conservation and resource extraction, high-value commodity chains, "green" precarious smallholder production, and agrarian change and development.